CAMBRIDGE LIBRARY COLLECTION

Books of enduring scholarly value

British and Irish History, Nineteenth Century

This series comprises contemporary or near-contemporary accounts of the political, economic and social history of the British Isles during the nineteenth century. It includes material on international diplomacy and trade, labour relations and the women's movement, developments in education and social welfare, religious emancipation, the justice system, and special events including the Great Exhibition of 1851.

A Journey through England and Scotland to the Hebrides in 1784

The French geologist Barthélemy Faujas de Saint-Fond (1741–1819) abandoned the legal profession to pursue studies in natural history, working at the museum of natural history in Paris and as royal commissioner of mines. His enthusiam for geology took him in 1784 to Britain, to investigate the basalt formations on the Hebridean island of Staffa described by Sir Joseph Banks in Pennant's *Tour in Scotland* (also reissued in this series). His subsequent account was published in France in 1797, and first translated into English in an abridged form in 1814. This two-volume annotated translation by the well-known geologist Sir Archibald Geikie (1835–1924), prefaced by a short biography of Faujas, was published in 1907. The work is interesting for its social as well as its geological observations. Volume 1 describes life in scientific circles in London, before recounting Faujas' journey to the Highlands of Scotland via Edinburgh and Glasgow.

Cambridge University Press has long been a pioneer in the reissuing of out-of-print titles from its own backlist, producing digital reprints of books that are still sought after by scholars and students but could not be reprinted economically using traditional technology. The Cambridge Library Collection extends this activity to a wider range of books which are still of importance to researchers and professionals, either for the source material they contain, or as landmarks in the history of their academic discipline.

Drawing from the world-renowned collections in the Cambridge University Library and other partner libraries, and guided by the advice of experts in each subject area, Cambridge University Press is using state-of-the-art scanning machines in its own Printing House to capture the content of each book selected for inclusion. The files are processed to give a consistently clear, crisp image, and the books finished to the high quality standard for which the Press is recognised around the world. The latest print-on-demand technology ensures that the books will remain available indefinitely, and that orders for single or multiple copies can quickly be supplied.

The Cambridge Library Collection brings back to life books of enduring scholarly value (including out-of-copyright works originally issued by other publishers) across a wide range of disciplines in the humanities and social sciences and in science and technology.

A Journey through England and Scotland to the Hebrides in 1784

A Revised Edition of the English Translation

VOLUME 1

BARTHÉLEMY FAUJAS DE SAINT-FOND
EDITED AND TRANSLATED BY
ARCHIBALD GEIKIE

CAMBRIDGE
UNIVERSITY PRESS

University Printing House, Cambridge, CB2 8BS, United Kingdom

Cambridge University Press is part of the University of Cambridge.
It furthers the University's mission by disseminating knowledge in the pursuit of education, learning and research at the highest international levels of excellence.

www.cambridge.org
Information on this title: www.cambridge.org/9781108071567

This edition first published 1907
This digitally printed version 2014

ISBN 978-1-108-07156-7 Paperback

A JOURNEY THROUGH ENGLAND AND SCOTLAND TO THE HEBRIDES IN 1784

IN TWO VOLUMES

VOLUME I

Impression, Four hundred and fifty copies.

A JOURNEY THROUGH ENGLAND AND SCOTLAND TO THE HEBRIDES IN 1784

BY

B. FAUJAS DE SAINT FOND

A REVISED EDITION OF THE ENGLISH TRANSLATION
EDITED, WITH NOTES
AND A MEMOIR OF THE AUTHOR

BY

SIR ARCHIBALD GEIKIE, D.C.L., SEC.R.S.
CORRESPONDENT OF THE INSTITUTE OF FRANCE

VOLUME ONE

GLASGOW: HUGH HOPKINS
1907

PREFACE

THE *Voyage en Angleterre, en Écosse et aux Îles Hébrides*, by Faujas de Saint Fond, was published at Paris in 1797 in two small octavo volumes, and also in an "édition de luxe" with the octavo pages printed on paper of quarto form. An English version of the work, by an anonymous translator, appeared at London two years later in two octavo volumes, like the original French edition. A greatly abridged form of this version was published in 1814 as part of the fifth volume of Mavor's *British Tourists*, with a prefatory statement in which it was remarked that the work had "already become scarce."

In the course of my studies I have had frequent occasion to refer to Faujas' *Voyage*, and have found that it is now but little known in this country. Last year at the meeting of the Franco-Scottish Society held in Aberdeen, I took occasion to make the book the subject of an address on the impressions of a French traveller in this country towards the end of the eighteenth century. Soon after the appearance of the press-notices of this meeting, I received from Mr Hopkins, publisher, Glasgow, an intimation that he had been for some time thinking of issuing a page for page reprint of the English translation of Faujas' work, and he asked whether I would be willing to undertake to edit the book. I consented to do so, but soon found that this translation required a good deal of correction, and that a verbatim reprint of it was not desirable. I have carefully revised it, and have added notes, chiefly to the scientific parts of the book, where the author's language seemed to stand in need of explanation or correction. All such editorial additions are placed within square brackets. I have likewise prefixed a Memoir giving a brief outline of the life of Faujas de Saint Fond and an appreciation of his geological observations in this country.

The plates have been reproduced by Mr Annan, of Glasgow, with his accustomed photographic skill. My friend Professor Lacroix, of the Institute of France, has been so good as to procure for me a photograph of the engraved likeness of the author preserved in the collection of the Muséum d'Histoire Naturelle, Paris, and from this photograph the portrait prefixed to this present volume has been prepared by Mr Annan.

As an intelligent and often racy narrative giving the impressions and experiences of a cultivated foreigner in this country in 1784 the *Journey* of Faujas de Saint Fond deserves not to be forgotten.

ARCH. GEIKIE.

Shepherd's Down, Haslemere,
30th March 1907.

CONTENTS

CHAPTER VII

CHAPTER VIII

CHAPTER IX

CHAPTER X

CHAPTER XI

CHAPTER XII

CHAPTER XIII

PLATES

MEMOIR OF THE AUTHOR

THE author of this *Journey*, Barthélemy Faujas de Saint Fond, was born on the 17th of May 1741 at the old town of Montélimart, in the valley of the Rhone. He came of a good family, which possessed the lands of Saint Fond in Dauphiné, whence they took the territorial part of their name. Having received his early education at the Jesuit College of Lyons, he afterwards studied law at Grenoble, where he passed as avocat. When only twenty-four years of age he became president of the senéchal court. But he would seem to have developed in his young days such an overmastering passion for natural history pursuits that when he came under the strong personal attraction of Buffon, his resolution was definitely taken to abandon the career for which he had been intended, and to devote himself to one that promised to be greatly more congenial. Buffon, the most distinguished and most influential naturalist of the day in France, probably had little difficulty in procuring for the young lawyer the appointment of Assistant Naturalist at the Muséum d'Histoire Naturelle at Paris, with the modest salary of 6000 francs or some £240 a year. Subsequently Faujas became Royal Commissioner of Mines, with an additional stipend of 4000 francs. At a still later period he received the appointment of Professor of Geology, which he held up to the time of his retirement in the year 1818.

In those days the range of the natural sciences was not so wide as to prevent an intelligent and industrious student from gaining a practical acquaintance with most of their branches and keeping himself abreast of their progress. The naturalist had not then become a specialist, confining most of his attention to one department, or even to a limited section of one department of investigation. He might be, and often was, at once an accomplished zoologist, a good botanist, a fair mineralogist, with a more or less detailed acquaintance with the physics and chemistry of the time. This wide breadth of acquirement, which was then so common, kept the whole brotherhood of students of nature in closer touch with each other than has become possible in our modern age of ever-increasing differentiation.

Faujas de Saint Fond was a naturalist of this broad type. While the study of rocks and minerals claimed his most continuous and absorbing attention, every object in the outer world had an attraction for him. He did not limit his enquiries to the different branches of natural history, but pushed them into the domains of chemistry and physics. Thus he was chemist enough to be able to make an analysis for himself of any particular rock or mineral that he met with. He took, moreover, so keen an interest in the then infant subject of aerial navigation that he was induced to enter upon the study of inflammable gas and of the varying density of different layers of the atmosphere, and he published in 1783-4 a treatise, in two octavo volumes, on balloons, wherein he discussed the art of making and the methods of steering them.

With all this varied scientific accomplishment he possessed an eminently practical mind, and seems never to have lost an opportunity of turning his knowledge to account for the furtherance of the arts and industries. He had a love for all sorts of machinery, and delighted to watch processes of manufacture, making careful notes of every novelty in idea or in method. And as he was an eminently patriotic Frenchman, he was always on the outlook, when travelling abroad, for improvements in the industrial arts which might be profitably introduced into his own country. He was justly proud of one of his achievements in the application of scientific knowledge to useful purposes. In 1775, while rambling through the district of the Velay, he was struck by the resemblance of a certain material in the hill of Chenavary to some of the volcanic deposits which, in Italy and elsewhere, have from time immemorial been used for making the best kinds of cement. At his own expense, he had the deposit opened up, and having himself analysed it chemically and found it to agree closely in composition with the famous Italian pozzuolano, he brought it to the notice of the government of the day, and in the end he had the satisfaction of seeing it used by the authorities in the construction of the harbour of Toulon and other public works.

From the side of the Rhone opposite to his native district the mountains of the Vivarais stretch westwards and northwards into the marvellously interesting and picturesque volcanic region of central France. To this ground Faujas was attracted in his youth, and it was doubtless there that he formed that passion for the study of volcanic rocks which became the dominant feature of his scientific career and now gives him his place of eminence among the early leaders of geology in France. In those days it was rare to see any man wandering among the hills with a hammer in his hand and a bag slung across his shoulder. Faujas tells us that, accoutred in this manner, he attracted the notice of two distinct types of onlookers. In the small towns, his critics of superior discernment used to ask him of what use those pursuits of his could be; what good could come of gathering together such quantities of stones which after all might be collected anywhere. But the peasants in the valleys and among the puys had more common sense than to put such questions. They felt sure that no one would take the trouble to pick up and preserve what was really worthless. When they saw a man so engaged, they were certain that he must be in search of some valuable mine. They would, therefore, watch attentively his movements, and as soon as he left the place, they would repair to it and procure an abundant store of what they had seen him collecting. They would then carefully guard it until an opportunity offered of visiting the nearest little town, where they would carry the treasure mysteriously to some jeweller who, to their surprise and disgust, would only laugh at them.

There was now and then, too, a spice of personal danger to the enterprising geologist who came unknown among an ignorant and suspicious peasantry. He might be taken for a government inspector sent to survey the ground preparatory to the imposition of fresh taxes, or a spy from the seigneur employed on some equally sinister errand. Thus in 1775 Guettard, the famous observer who first recognised the volcanic origin of the Puys of Auvergne, visited Faujas in Dauphine and was taken by him to see some of the geological wonders of the Vivarais. Halting for the night at Vals, near some fine basalt colonnades, they put up at a miserable public-house. Lodged in a room over the kitchen, Faujas through chinks in the floor could hear, on the one hand, the loud and angry voices of a company of muleteers upbraiding the mistress of the house for harbouring two such suspicious visitors, and on the other, the soothing efforts of their adroit hostess, who kept plying the men with wine until they were mellowed into an agreement to postpone further action until the morning. When the muleteers discovered on getting up next day that the travellers were only preparing boxes in which to send off bits of stone, their suspicions were allayed, and after numerous facetious and absurd questions they all parted good friends.

As the result of all these years of wandering in the Vivarais and Velay, Faujas prepared and published in 1778 a magnificent folio volume on the extinct volcanoes of that region, adorned with many excellent plates, in which the volcanic nature of the hills was admirably revealed. This work appeared at a time when an active controversy was in progress in Europe as to whether basalt is a product of volcanic action or has been formed as a chemical precipitate in water. The observations of Guettard and Desmarest among the hills of Auvergne ought to have settled this dispute, and had there remained any doubt on the subject the great volume of Faujas should have removed it. But the warfare continued for many years, and did not finally die out until these early protagonists had been laid in their graves. Faujas will, however, be always held in honour for his services in the settlement of the question, and his splendid volume is now one of the prized classics of geological literature. Though only thirty-seven years of age when his great work appeared he was already in active correspondence with some of the foremost naturalists of France, and he printed at the end of the book an interesting series of their letters to him. Those from Buffon and Dolomieu are more particularly to be noted.

The spirit in which he carried on the researches that culminated in his folio may be judged from the following sentences taken from the Preface to the work. "Experiments in laboratories ought always to be subordinated to examination in the field. It is there that the careful and enlightened observer gains the preliminary knowledge which gives firmness to his steps and leads him on to discovery. There is a chemistry of Nature far higher than that of art; experienced eyes can perceive it, follow its traces and distinguish its processes and its effects. Nature, which never reckons with time, nor hurries anything, works by slow, insensible, but uninterrupted stages, and always in accordance with laws as simple as herself."

It was six years after the publication of this volume that he undertook the journey described in the present work. The island of Staffa, with its marvellous columnar cliffs and caves, had been first brought to notice in 1772 by Sir Joseph Banks' account of it, which, together with his plates of the scenery, was published in Pennant's *Tour in Scotland*. Faujas' curiosity was evidently keenly excited by this publication, and he determined to undertake what at that time was by no means an easy journey, and to make his way across England and Scotland to the now famous isle set in the western sea. It is this journey which is described by him in the following chapters.

He came to England with good introductions, and was received everywhere with much cordiality. As stated on his title-page his object was to take note of the state of science in this country, more particularly with reference to natural history, to make himself acquainted with the condition of various arts and industries and to observe the manners and customs of the people. His account of his visit thus comprises a miscellaneous series of records pertaining to scientific subjects, more especially to the geology of the districts through which he passed, mingled with notices of manufactures, scenery, domestic habits, machinery, meetings and conversations with eminent men, and any other topics which interested so shrewd and observant a traveller. Some portions of his narrative are valuable for the pleasant light which they cast on the more intimate life of some of the famous Englishmen and Scotsmen of the latter part of the eighteenth century, such as Joseph Banks, William Herschel, James Hutton, Adam Smith, James Watt, Joseph Priestley, and many more. He gives a graphic description of a dinner of the Royal Society Club in London, and, indeed, generally takes pains to note the prandial and convivial customs which he met with throughout his tour. Not less amusing is his account of the bagpipe competition in Edinburgh to which he was taken by Adam Smith, who in subjecting his French guest to the ordeal, seems to have combined the qualities of a scientific experiment with those of a practical joke. The state of the roads throughout the country, the character of the inns, the distinctive features of the centres of industry, the conditions of agriculture, the breeds of sheep, are all commented upon as he travels along, and his remarks afford material for an interesting comparison with the state of things at the present day after the lapse of nearly a century and a quarter.

Faujas de Saint Fond was above all a geologist, and the most important parts of his *Journey* are perhaps those which record his geological observations. He was the first geologist who ever set foot on Staffa and described it as a volcanic island. The notes which he has left regarding the nature of the rocks examined by him in the course of his traverses of the country have all the interest of those of a pioneer. It is true that they faithfully reflect the imperfect state of the science at the time, and exhibit the crude notions entertained as to the origin and classification of rocks. But copied out of his journal as they were written from day to day, they seem to admit us into the companionship of a field-geologist of that early time, showing how he worked and how he applied the nomenclature then in vogue to the fresh materials which he collected.

The inaccuracies of his determinations must be frankly acknowledged. Some of them, however, were only part of the imperfect lithology of his day, and others may have been due to the necessarily hasty character of observations made in the course of daily travel, which might not improbably have been in some cases corrected from a more leisurely examination of his specimens had not his collection been lost by shipwreck on its way to France.

Although his reputation as a volcanist was well-established at the time of his visit to this country, it is difficult to make out from his *Journal* what were his real opinions as to the nature and origin of volcanic action. From his constant use of such phrases as "subterranean conflagrations," "combustions," and "fires," the impression is conveyed that these are not mere popular language, but that he then really held the old notion, adopted also by Werner and his followers, that volcanoes derive their energy from the combustion of beds of coal or other inflammable material within the earth. This impression is strengthened by occasional references to the kinds of rock by the melting down of which different lavas have been produced, and to instances where, though the rock had been "volcanised in place," it retained traces of its primitive condition.

While he stood up as a stout upholder of the volcanic nature of basalt, Faujas was as convinced a Neptunist as Werner himself with regard to the aqueous origin of the so-called "trap-rocks." It is curious to see how this prepossession leads him to explain away or disregard evidence which he himself detected in the field and which he admits to afford apparently a presumption that the toadstone of Derbyshire is an igneous rock. He controverted the published opinions of Whitehurst on this subject, and missed the opportunity of turning his experience in volcanic districts to account by demonstrating, what has since been clearly ascertained, that the toadstones afford an interesting record of a volcanic episode in the geological history of Britain when streams of lava and showers of volcanic ashes were discharged over the floor of the Carboniferous Limestone sea.

While in this way he refused to recognise the truly volcanic nature of certain rocks which he confidently maintained to be aqueous deposits, he identified as lavas others which are unquestionably mere hardened marine sand and mud. In the southern uplands of Scotland and in the Lake district of the north of England he took the Silurian grits, sandstones and shales for volcanic masses,—lavas and basalts which sometimes "imitated the fissile structure of slate."

But when all these errors and limitations are fully recognised the merit of the geological observations of Faujas in Scotland must be ungrudgingly acknowledged. He saw, and was the first to see, the volcanic nature of the basalt in the Inner Hebrides, of the terraced hills of Lorne, of the Ochil Hills near Perth, of the rocks on the coast of Fife at Kinghorn, and of Arthur's Seat at Edinburgh. It was an important event in the history of science at a time when the battle was raging over the origin of basalt, that an experienced observer from the cones and craters and lava-streams of the Vivarais and Velay should visit this island and recognise here some of the same proofs of former volcanic activity with which he was familiar in the south of France. But the importance of this recognition failed to meet with appreciation in Scotland. Robert Jameson, who followed Faujas' steps along part of the west of Scotland, published in 1800 his *Mineralogy of the Scottish Isles*, in which he indulged in sneers at the volcanist doctrines, and almost contemptuously rejected the conclusions of his French predecessor, whom he cites by name. "I do not hesitate a moment," he wrote, "in saying that, in my opinion, there is not in all Scotland the vestige of a volcano," and he elsewhere expresses the hope that at a place which had been "said to be the crater of a volcano, probably there may be still sufficient heat to revive the drooping spirits of some forlorn fire-philosopher as he wanders through this cold bleak country." * This zealous disciple of the Freiberg School was appointed in 1804 Professor of Natural History at the University of Edinburgh, which thereupon became, and continued for many years to be, the centre of Wernerian propagandism in Britain, while the early observations of Faujas in 1784 gradually passed out of mind.

Though Faujas' visit to Britain took place in 1784 his account of it did not appear until 1797. Explaining in a Postscript that his work was in the press in the second year of the Revolution, but that the social convulsions of the time put a stop to its progress, he makes apologetic reference to the occasional footnotes in which he expresses his feelings

* *Op. cit.*, vol. i. pp. 5, 86.

as to the irreparable loss inflicted on France by the tragic deaths of so many of its most illustrious citizens. What his own injuries were, which he says he will gladly forget, does not appear. In some respects, however, the government of the Revolution recognised his merits. Not only did he retain his official post, but when in 1793 the National Convention reorganised and enlarged the staff of the Museum, he was given the Professorship of Geology. It was at this same time also and at the same Institution that Lamarck received the Chair of Invertebrate Zoology. Faujas had spent largely out of his private fortune in augmenting the collections in the Cabinet of Natural History, and the Council of Five Hundred voted him a sum of 25,000 francs towards reimbursing him for his outlays. He continued to make geological excursions and to contribute papers on various subjects to the *Annales du Muséum* and other journals. He edited the works of Bernard Palissy, famous not only as a potter but as the first Frenchman to perceive the geological bearing of fossil shells. He supplied notes to the French translation of Spallanzani's *Viaggi alle due*

Sicilie (1792-3). At the age of 72 he almost entirely recast his treatise on Trap-rocks, and published a second edition of the work. For six years longer he remained at his post in the Museum. Advancing years, however, induced him at last in 1818 to resign and to quit Paris for his estate of Saint Fond. There in the home of his childhood he spent the last months of his long and active life, dying the following year in the 79th year of his age.

A JOURNEY

THROUGH

ENGLAND AND SCOTLAND

TO THE

HEBRIDES

CHAPTER I

LONDON

Sir Joseph Banks.—Doctor Whitehurst.—Cavallo.—Doctor Letsom.—Sheldon.—Royal Society.—British Museum, &c.

I DO not intend to entertain the reader, under this head, with observations on the extent, the beauty, or the immense population, of the city of London: * these have been fully treated of by others.

* Arthur Young, in his Travels in France, during the years 1787, 88, 89, and 90, a work full of new views and instructive facts respecting the population, commerce, and agriculture of France, compared to that of England, in speaking of the extent and population of the city of London, asserts that London is so superior to Paris, as to render all comparison ridiculous; and that he believes that, without any exaggeration, this city alone is equal to Paris, Lyons, Bourdeaux, and Marseilles, as might be proved from the accounts of the population, wealth, and commerce, of all these places. However much I esteem

Natural history, the sciences, the arts, and some objects of commercial economy, will more particularly engage my attention. I proceed to the subject.

Sir Joseph Banks *

THE house of this celebrated traveller, who is President of the Royal Society of London, is the rendezvous of those who cultivate the sciences; and foreigners are always received there with politeness and affability. They assemble every morning in one of the apartments of an extensive library, which consists entirely of books on natural history, and is the completest of its kind in existence. There all the journals and public papers, relative to the sciences, are to be found; and there the members of the party communicate to each other such new discoveries, as they are informed of by their re-

Arthur Young, I am not entirely of his opinion. London is, doubtless, more extensive and populous than Paris; but not in such disproportion.

* [Sir Joseph Banks (1743-1820) accompanied Cook in his voyage round the world (1768-1771); afterwards visited Iceland; inherited a large fortune, which he munificently spent in promoting science; elected F.R.S. 1766; became President of the Royal Society in 1778, and remained in that office until his death. His house was the chief centre of scientific activity and intercourse in London.]

spective correspondents, or which are transmitted by the learned foreigners who visit London, and who are all admitted into this society.

A friendly breakfast of tea or coffee maintains that tone of ease and fraternity, which ought universally to prevail among men of science and of letters. They would, in general, become more sociable, and more intimately allied, if accustomed to meet each other frequently; especially if they enjoyed, as in this city, a rallying point, presenting the charms of agreeable society in conjunction with those pure pleasures, which every thing connected with knowledge and instruction must always afford.

Sir Joseph Banks has rendered important services to the sciences, by his distant voyages, undertaken for the purposes of discovery; he benefits them daily by the collections which have been the result of those voyages, particularly that of botany, the branch of natural history in which he is most interested. His fortune also enables him to carry on an extensive correspondence with every part of the globe; and his kind and communicative disposition admits others to participate in the fruit of all his new discoveries.

This friend of the sciences has hitherto published only the Collection of William Houston,* containing a number of American plants, and Kaempfer's Collection of the Plants of Japan; but he has been long occupied in having the ablest artists to draw and engrave the plates of a superb collection of the plants of the South Sea; a valuable work, which is expected with impatience, and which will be worthy of the reputation of its author.

Sir Joseph Banks has made some valuable acquisitions in botany, and has thence become the guardian of several herbals made by naturalists of great reputation. Had it not been for his attention and fortune, these collections of plants might have been dispersed, or, perhaps, lost by negligence of heirs; whereas, united as they now are, in one repository, they are easily accessible to such as desire to consult them.

It is with this view that he purchased from

* Reliquæ Houstonianæ, seu Plantarum in America Meridionali a Gulielmo Houston collectarum, icones manu propria incisæ, in bibliotheca Joseph Banks asservatæ. Londini, 1781, in 4to, fig.

Mr Dick, minister of the gospel at Bodligen in Switzerland, an herbal of the celebrated Haller, which comprizes the collections of plants made by the two brothers, Bauhins.

He has likewise acquired the plants of French Guiana, by Fusée Aublet, the herbal of Jacquin, and several others formed by those naturalists who have illustrated botanical science.

At the time I had the pleasure of seeing Sir Joseph, he had just received from China a package, in which there was a box, containing a stony material, reduced to powder, such as lapidaries employ in cutting and grinding hard stones of the nature of rock-crystal, which the Chinese make use of as ornaments in their dress and furniture. The same box inclosed a small bag, containing some fragments of the stone from which this powder is made.

Europeans, in general, are acquainted only with two substances fit for sawing and polishing hard stones—the powder of diamond, for cutting the diamond itself and other gems; and that of emery, for jaspers, agates, rock-crystals, &c.

Several years before the transmitting of

that parcel, Doctor Lind, being at Canton, employed himself in enquiries respecting the Chinese arts; and he spared neither pains nor expense to obtain information on that subject, notwithstanding the difficulties which he had to surmount.

He brought away with him some of the same powder, and one of the bows used by the lapidaries of China. It differs from that employed by Europeans, in having its string made of double wire, that is, of two pieces of wire intertwisted; whereas ours consists of a single wire only. The Chinese method deserves to be made trial of by our artists. Doctor Lind did not neglect to procure some specimens of the stone which supplies the powder for cutting; and it so happened, that almost all the fragments which he obtained, were crystallized. The largest of these crystals was sent by Doctor Lind, in 1782, to Doctor Solander, who accompanied Sir Joseph Banks in his voyage round the world. This celebrated naturalist died a short time afterwards; and Mr Wolfe, an able chemist, who purchased it at the sale of Solander's effects, very obligingly made me a present of it at Paris, in the year 1783, where I then had the pleasure of seeing him.* This is the same crystal which I lent to Mr Brisson to ascertain its specific gravity, and which he has mentioned in his work on the Specific Gravity of Stones and Minerals.

It was by following the directions of Doctor Lind, that Sir Joseph Banks endeavoured to procure the powder and specimens of the stone from China, and that he succeeded in getting home the box, containing several pieces of it. Sir Joseph offered me my choice of such as struck my fancy, and gave me a sufficient quantity of the powder to enable me to make various experiments with it at Paris.

What attracted me most in the choice which Sir Joseph allowed me, were some pieces still adhering to portions of the stones in which this *adamantine spar* † (as Doctor Lind calls it) is found. By their means, indeed, I was able to determine the nature of the compound substance that contains this stone, which, at first view, and from its lamellated structure, resembles a felspar. I think it is beyond doubt, that this substance

* I have given this fine crystal to the National Museum of Natural History at Paris.

† [One of the old names for Corundum.]

is contained in a species of granite, in which it exists in the form of crystals, of a greater or smaller size, and more or less perfect. These crystals are opaque, of a greenish-brown, and sometimes of a greyish-white colour, and they attract the magnet. They ought, therefore, to be sought for in granite rocks, where it is probable they may be found; and considering their usefulness in the art of cutting hard stones, it would be a desirable thing to discover them in the granites of France; we might then dispense with using emery, which we bring from abroad.* Doctor Lind gave

* The following is a list of the specimens which I selected from those sent to Sir Joseph Banks, and which I gave to the Museum of Natural History at Paris, together with the large crystal presented to me by Mr Wolfe.

1. Adamantine spar, of a brown colour, with rhomboidal cross-sections in some parts.

2. Idem, adhering to lamellar mica, of a bright silver colour.

3. Idem, with a glossy black mica.

4. Idem, with greenish felspar, black mica, and some specks of black schorl.

5. Adamantine spar, adhering to a greenish steatite, hard but smooth to the touch, and yielding an unctuous powder.

6. Idem, with white steatite, soft and slightly micaceous.

7. With rose-coloured felspar.

8. With ferruginous pyrites of a cubical form.

9. Crystal, mixed with adamantine spar, having on one of its sides an impression of a cubical pyrites.

10. Adamantine spar, in large irregular particles, adhering to like irregular particles of a reddish coloured quartz, and micaceous iron in small laminæ.

the Chinese stone the name of *adamantine spar*, because its lamellated texture, its crystallization, and its appearance on being broken, seemed to class it among the *spars*.

The generic term *spar*, which we have from the Germans, our first masters in mineralogy, though itself without meaning, ought not to be rejected, as the partizans of the new nomenclatures would like. On the

It cannot be any longer doubful, that the adamantine spar exists in a compound rock. Lametherie mentions the adamantine spar in his Notes on the sciagraphy of Bergmann, p. 271, vol. 1, and, in imitation of Klaproth, gives it the name of Corrindon. This learned and esteemed friend, who seeks only for truth in all his enquiries, will not be displeased that I should correct a trifling error which he has committed on the subject of this stone, where he says, in speaking of its hardness, "that it equals only that of the rock-crystal at most, for rock-crystal cuts it sooner than it cuts the crystal; its hardness may, therefore, be estimated like that of crystal at eleven." The following is a decisive answer upon this subject: it is a detailed account of the experiments which I caused the son of M. Fontaine, one of the ablest lapidaries in Paris, to make in the presence of M. Hoppe, a German, who is deeply versed in the knowledge of precious stones. This subject is of sufficient importance, from its novelty and utility, to justify my introducing in this place the result of these experiments.

"M. Faujas (Saint Fond) having sent to M. Hoppe a certain quantity of adamantine spar, to make comparative trials of it with emery, the latter applied for that purpose to M. Fontaine the son, whose talents are well known to every amateur, and who agreed, in the most obliging manner, to satisfy the wishes of M. Faujas.

"The principal operations of the lapidary on precious stones

contrary, it is because we perceive no particular meaning in the root of that term that it should be considered as a good one, when used in conjunction with an epithet; as, for example *calcareous* spar, *heavy* spar, *cubic* or *phosphoric* spar.

Klaproth, who has analysed the adamantine spar like an able chemist, has given it, for what reason I know not, the name of Corrindon; and having discovered in that

being to cut, to drill, and to shape on the wheel, M. Fontaine employed the adamantine spar instead of emery in all these operations, and, that he might obtain more certain results, with instruments which had not been in use before.

"Lapidaries cut and drill, in general, with powder of diamond; to this the adamantine spar was much inferior, but its effect was pretty conspicuous, and superior to that of the emery. M. Fontaine compared it to that which the powder of ruby, sapphire, or oriental topaz, might produce.

"He then employed the powder of adamantine spar in grinding and shaping on the wheel; and its effect surpassed, in a degree beyond all expectation, that of the other substance. The result of a great number of experiments was, that the lapidary's wheel retains very well the adamantine spar, that it requires only a fourth part of what would be necessary of emery to render it perfectly fit for grinding, and that one-half of the time is saved in the operation. In may be proper to add, that the adamantine spar prepares stones better for receiving a polish than emery, because the first produces a finer grain. M. Hoppe having been present at all these operations, took the above minute of the result of the comparative experiments, and signs it with M. Fontaine.

"*Paris, May* 30, 1789. HOPPE. FONTAINE the son, lapidary."

stone an earth peculiar to it, and entirely *sui generis*, united with quartz, iron and nickel, he has thought fit to call it the *corindonian* earth. If the word *corindon* were Chinese,* I should be the first to adopt it, and to advise naturalists to preserve it, were it only to prove that the stone, and the use to which it is applied, have been transmitted to us by the Chinese. But this word has so little resemblance to the language of that people, that it must be presumed to be of European manufacture.

This rage for coining new words has seduced M. Haüy, a very able naturalist, to create one of a compound kind, in imitation of some chemists, who have endeavoured to compress the principles of science into its terms. He has, accordingly, called the adamantine spar *leïaste*, signifying *lævigator* (polisher), on account of the use, as he says, to which it is applied, of polishing stones. But M. Haüy should have considered, before he formed a name taken from the properties of the thing itself, that the powder of adamantine spar is employed in cutting and sawing stones, and not in polishing them.

* [It is the Hindustani word *Kurund.*]

It is to the *putty* and the *ruddle* of England, which are used for giving to stones their polish and lustre, that the name of *leïaste* would be most applicable. Let us, then, preserve to this stone that of *adamantine spar*, however imperfect it may be, until we have ascertained the name which it bears in China. In the meantime it will continue to be known, that Doctor Lind was the first who made us acquainted with it, and with its use among a people whose high antiquity and steady application have given birth to a number of arts and processes, which might be easily proved to be still unknown to Europeans.

The reader will, I hope, excuse the details into which I have entered respecting this stone; as it is an object still new in natural history, and may be of advantage to the art of the lapidary. Since it is presumable that adamantine spar does not exist exclusively in the granites, or among the porphyries of China, I conceived that these considerations merited the attention of men of science.*

* It would be too long, and rather misplaced, to mention here the different analyses which have been made of this stone in France and Germany. M. de Bournon thought that he perceived some traces of it in a granite which had been brought from the Forêt, near Montbrison; and others conceived that

The package sent to Sir Joseph Banks contained another object not less worthy of attention, as it interests a more numerous class of men—that of those who are engaged in agriculture: it was a quantity of the seed of a species of hemp of a superior quality to that cultivated in Europe.

In the year 1781, Mr Elliot, who resided some time at Canton, gave Mr Fitzgerald from thirty to forty grains of this hemp seed. It was not sown until the 4th of June, which was about a month too late; but, notwithstanding this circumstance, and the dryness of that season, the greater number of the plants rose to the height of fourteen feet, and the stems of several were so large as to measure seven English inches in circumference.

These promising plants blossomed; but on the approach of cold weather, they perished without yielding any seed. "There are," says Mr Fitzgerald, "from thirty to forty lateral branches on a plant; they are set off in pairs, one on each side of the stem, pointing horizontally; the others, at about

they had made a like discovery in other quarters. I intend, as soon as my avocations will permit, to investigate this subject completely in a separate essay.

five or six inches distance from them, pointing in different directions, and so on to the top; the bottom branches of some measuring more than five feet, the others decreasing gradually in length towards the top, so as to form a beautiful cone when in flower.

"After having steeped them in the ordinary manner, on trying whether the hemp could be easily separated from the woody part, I was agreeably surprised to find, that on peeling a few inches longitudinally from the root, the whole rind, from the bottom to the top, not only of the stem, but also of all the lateral branches, stripped off cleanly, without any one of them breaking: the toughness of the hemp seemed to be extraordinary. . . .

"The rough hemp that had been peeled from thirty-two plants, when thoroughly dried, weighed three pounds and a quarter; but I do not think it had come to full maturity, though I can hardly doubt but the plants would have come to perfection, if the seed had been sown in the proper season. The summer was remarkably dry, notwithstanding which, although the situation they were placed in was very warm, and the ground not rich, I found, on measuring the plants at different times, that they had grown nearly eleven inches *per* week."

Such is the substance of the paper communicated by Mr Fitzgerald to Sir Joseph Banks, which was read to the Royal Society of London, on the 17th of January, 1782, and printed among the *Philosophical Transactions* of that year. I conceived that my readers would peruse, with some interest, the history and result of the first experiment, though an incomplete one, which was made in Europe on the hemp of China.

The desire of acquiring for England an object so important for the navy which constitutes her power, engaged the attention of the government, and of those who wished to enrich agriculture with this new production. It was known that the exportation of this seed was severely prohibited in China. This prohibition, however, served only as a fresh stimulus to zeal and emulation; and, on the first return from the voyage to China, a vessel brought home about a pound of it, concealed in the box of natural history, addressed to Sir Joseph Banks, which contained the adamantine spar, and other curious objects. The President of the Royal Society was pleased to divide two ounces between Broussonet, who was then in London, and myself, in order to make trial of it in France, particularly in the southern departments, where both of us had estates. There was reason to hope that the seed would come to maturity in that climate, and that a second supply might be got there, in case it should not succeed in England. We received this valuable present with suitable acknowledgments; and I resolved, on my return to Paris, to distribute some of it to able cultivators, and to reserve the remainder for the south of France.*

* On my return to Paris, my first business was to give some grains to M. de Malesherbes, M. de Rosambo, his son-in-law, M. de Trudaine the elder, M. Boutin, M. de Lavoisier, M. Hell for Alsace, Varenne de Fenille for Bresse, M. de Buffon for his estate of Montbard, in Burgundy, M. Thoin for the Botanical Garden of Paris, and to M. Moral, physician at Montélimar, who employs himself in agricultural pursuits. I reserved about fifty grains for the subsequent year, intending myself to direct the sowing of it, and to attend to its progress in the department of Drôme.

The hemp of China succeeded every where, even beyond expectation. That in the Botanical Garden of Paris was visited as a curiosity; in the month of August it was fourteen feet high, and from fifteen to sixteen in September. M. Hell wrote to me, from Alsace, that the stems of the twenty grains which he had sown, were seventeen feet high on the 20th of September, and that their lateral branches were so vigorous, and occupied so much space, that all the plants seemed rather trees than annual vegetables. This was the case also at

Whitehurst

Benjamin Franklin had the goodness to give me a letter of recommendation to his old and estimable friend Whitehurst, who has illustrated, in the most ample detail, the wonderful and singular structure of the mountains of Derbyshire, in a work entitled "*Inquiry into the original State and Forma-*

Malesherbes, Montigny, in Burgundy, Bresse, and the other places. Every where male and female plants, of a promising blossom, made their appearance; but from Bourg to Paris, from Paris to the bailiwick of Landser, in Alsace, not one of them yielded good seed; the cold having prevented them from ripening. I was also informed, that all that had been sown in England had the same fate. This species of hemp was every where found superior to the common sort, in strength, in its silky texture, and in the length of its fibres. M. Thoin ventured to raise a few plants in turf, for the purpose of being put into pots, and placed in a green-house; so anxious was he to prevent this gigantic hemp from being lost. The seed acquired a certain degree of maturity; it was not very vigorous, but it was capable of reproduction. I had some hopes from the other places where it had been tried, and I particularly looked for an account of what had been sown in the vicinity of Montélimar. I learned, towards the end of September, that it had perfectly succeeded; that the dryness of the season prevented it from growing so high as it did elsewhere; but that the seed had acquired the proper ripeness, and that more than a pound of it had been gathered. I obtained the same success the following year from that which I kept in reserve, and which I sowed in the department of Drôme; the plants rose to the height of twelve feet. I have hitherto continued this

tion of the Earth, &c. by John Whitehurst. London, 1778." *

He was a native of Derbyshire [Cheshire], and resided a long time in the principal town of that county. His views and his meditations were thus directed to the face of a country truly extraordinary, and interesting to naturalists.

Whitehurst formed himself at an epoch when this science was not far advanced.

branch of cultivation with much success, and it is evident that the hemp of China produces excellent seed in the south of France. I have already given quantities of it to several persons at Paris, and other places. It is of very great importance to avoid sowing it near the common hemp. I wait for peace to repay my obligation to the English; for it is no more than just to return that which they have so generously lent us. I should have published a long time ago the result of the experiments which I made myself, and those of the different persons to whom I had given part of the seed; but I have been prevented, I must confess, by the sad recollection that of eleven persons to whom I gave some of the seed from China, and who, with an enthusiasm for the public interest, devoted their whole attention to its cultivation, eight have been dragged to the scaffold, without respect for names signalized by virtue and talents. Buffon was dead—they took his son—he was unrelentingly assassinated. . . .

* [John Whitehurst (1713-1788), born at Congleton, in Cheshire, son of a watchmaker; settled in Derby, where he gained so high a reputation as maker of philosophical instruments and a skilled and ingenious mechanician, that he received (1775) an appointment in connection with the Mint, and then settled in London. He became F.R.S. in 1779. See chap. xvii. vol. ii.]

But if he has committed errors (and who is free from them?) they are owing, not so much to the uncertain state of mineralogy at that period, as to a sort of religious awe and restraint which has often fettered the progress of men of genius, and compelled them to trace immediate connexions between the multiplied revolutions of the globe, and the oriental cosmogony, delivered in the books of Moses.

Whitehurst had, many years before, fixed his residence in London, where he had it more in his power to prosecute his studies, and to cultivate the society of men of learning. This venerable old man was very tenacious of a discovery which he thought he had made in Derbyshire, of a number of currents of basaltic lava crossing in various directions banks of calcareous stone, which they seemed, in many cases, to have moved from their original position. The whole of the descriptive part of his book is excellent, and remarkably accurate.

"You have," said he to me, "seen many volcanised countries; you have made a particular study of the different substances acted on by subterranean fire. I have not been able, from particular circumstances, to travel out of England, and to see volcanos in action; but I conceived that I discovered in the mountains of Derbyshire traces of subterranean combustion so evident, that I thought myself capable of building on that basis a system relative to the ancient state of the earth. I am anxious to have my observations confirmed or rejected by you.

"Were it not for a sick female relation in this place, who requires all my attention, I should quit every other employment to accompany you. I feel a great privation in losing this opportunity; but I will request you to be the bearer of a letter for a physician at the mineral wells of Buxton, who is a man of general information, and well acquainted with the places described in my book."

Manners of this affable complexion are calculated to render the sciences attractive, and, when one has the happiness to meet with such men, to procure esteem for those who cultivate them.

I found a remarkable resemblance between Mr Whitehurst and his friend, Benjamin

Franklin. His good-nature, his frankness, his admirable simplicity of manners, and mild philanthropy, engaged my attachment, and I visited him frequently. He had the goodness to introduce me to the acquaintance of several men of science, and to conduct me to the houses of the ablest artists of London, with whom he was particularly intimate, and who often came to consult him. He carried his attention so far, as to divide with me a part of the minerals and fossils which he had collected, and on which he supported his system. His cabinet was not very considerable, his collection being confined to the productions of Derbyshire; but an assemblage of objects, suitable to form the basis of a mineralogical description of a country, will always present much interest to the scientific traveller.

I promised Mr Whitehurst, that I should direct my particular attention to the mountains which he had described. He said I should oblige him by writing to him from Buxton or Matlock, as soon as I had visited that part of the country, and telling him, without any reserve, what I thought of the toadstone, and the other stones which he regarded as volcanic. All the specimens which I saw at his house convinced me, that he had fallen into the same error with Lammanon, who, in the Alps of Champsaur, mistook *traps* for lavas; but I waited to see the places where they had been collected, before expressing to him my opinion on the subject.* I acted in the same manner towards my industrious and adventurous friend Lammanon, who was convinced of his error a little before he set out on the voyage round the world with La Pérouse,† whose unhappy fate he shared.

* [Whitehurst had described the Derbyshire toadstone as "being as much a lava as that which flows from Hecla, Vesuvius, or Aetna" (*op. cit.* 2nd Edit., 1786, p. 185). In this opinion he was undoubtedly right, as has been abundantly proved since his time. But see chap. xvii. of vol. ii. of the present work.]

† Lammanon, who had great talents and a strong attachment to natural history, was so convinced of the existence of an extinct volcano, which he believed he had discovered in the mountain of Drouveire, one of the High Alps of Champsaur, in Dauphiny, that he made a drawing of the crater, the currents of lava, in a word, the complete topographical plan of these supposed remains of a subterranean fire in the Alps; where there exists not a single trace of a volcano. He transmitted to me from Turin, where he then was, the manuscript of the paper which he composed on this subject; and requested, in his letter, to know my opinion respecting the discovery. He sent to me, at the same time, a collection of the different substances which bore, according to him, the most evident marks of the action of volcanic fire. I had visited that mountain, in 1776, in company with the botanist Liotard, and I was already in possession of a

*Cavallo.**

Tiberius Cavallo is a native of Naples, but has been settled upwards of twenty years in London, where his chief occupation is natural philosophy, in which he is highly skilled: he possesses much information, and displays singular dexterity in performing the most

collection of the stones which are found there, and which, at first sight, appear to resemble lava. I answered Lammanon's paper, article by article, and traced a parallel between these stones and the *traps* of Sweden, of which I had a good collection. Lammanon, in his turn, refuted my objections in a very ingenious manner; and persisting in his opinion, on his return to Paris, printed at Cuchet's, bookseller, Rue et Hôtel Serpente, his paper, my letter, and his answer, with a plan of the places. This work was about to be published in 1784, when some learned Swedish mineralogists, to whom he communicated it, together with his collection, assured him that all those substances were varieties of *traps*, absolutely similar to those of their country. Lammanon, who fought only for truth, acknowledged his error: he did more; he suppressed and destroyed the whole impression of his book, excepting twelve copies, to all of which he annexed a printed paper, containing an honourable confession of his mistake, and the name of the person to whom he gave each of them. He was pleased to include me in the distribution. This book is a great rarity, as there are only twelve copies in existence, or more properly eleven, since the death of the author, who carried one with him on his voyage. The title of it is "*Mémoire Litho-Géologique sur la Vallée de Champsaur et la Montagne de Drouveire, dans le Haut-Dauphiné, par le Chevalier de Lammanon. Paris*, 1784, 8°. *avec une carte.*"

* [Born at Naples in 1749; settled in London, where he distinguished himself by his inventions and investigations in electrical science. He became a Fellow of the Royal Society in 1779.]

delicate experiments. He has carried electrometers to such a degree of perfection, as to make them sensible to quantities of electricity inappreciable by ordinary instruments. I saw at his house thermometers, upon which the slightest change in the temperature of the air acted in a very remarkable manner, and by means of which he has obtained divisions and graduations which had never been reached before. The tubes are extremely slender, and of a perfect bore, and the mercury is brought to the utmost degree of purity.

Cavallo has contributed greatly to that high degree of accuracy and perfection to which philosophical instruments have been carried in England. I must also do justice to the English artists; they enter upon their profession with much preliminary knowledge, resulting from a good education, which does honour to a country where the majority of the people are in a condition to estimate the merits of able artists, and to put an adequate value on the productions which come from their hands.

Cavallo was, at one time, much occupied with aerostatic globes [balloons]. He has translated into English, with notes and commentaries, all that has been written in France on that astonishing discovery, of which the progress has been interrupted only by the great expenses requisite for experiments on a large scale; respecting which, however, we can never arrive at any satisfactory results, without operating with vast machines, and employing the most extensive means. But the principle is known; and it is to be hoped that, sooner or later, it will be proceeded upon, and that some real advantage will be derived from a discovery, the value of which has not yet been generally enough felt.*

M. Cavallo shewed me a small, but simple and ingenious apparatus to procure ice quickly, even in the hottest time of the dog-days, by the simple process of evaporation.

* The French used an aerostatic globe of taffety, with great advantage, at the battle of Fleurus. The inflammable air introduced into it was obtained from a decomposition of water, by means of iron and a violent fire. It is the first time since men have fought against each other (and they have fought ever since they existed) that such a military machine was seen in the presence of an enemy: it was entrusted to the care of Coutelle, a man as intelligent as modest, who was captain of a company wholly destined to this service, and formed of young soldiers full of zeal and address. It was an agreeable sight to see them conduct, raise, lower, and manœuvre, in every direction, this moving observatory, which enabled the general-in-chief to contemplate, at one view, and to a great distance, all the dispositions and motions of the enemy.

This instrument consists of a small cylindrical tube of very thin glass, about four or five lines in diameter, and two and a half or three inches in length, open at one end, and shut at the other.

There is introduced into this tube, till it reaches the bottom, a wire of any metal, but very slender, and of a spiral form; water is then poured into it to the height of seven or eight lines. The metallic wire is only meant to draw out the ice, when it is formed. Things being in this state of preparation, one of those small glass syringes with which children amuse themselves is taken; its extremity or beak should be of the smallest capillary size, and it should have no piston. A quantity of well-rectified vitriolic ether is then poured into this kind of funnel, and the upper orifice is stopped with the thumb to prevent the evaporation, and to force the volatile liquid to escape by the small end.

The ether soon trickles in very small drops from this lengthened bill, which is held over the cylindrical tube, containing the water to be converted into ice. The ether is thus made to fall on the exterior surface of the tube, which is held in the left hand, and turned about with the fingers, till the whole is wetted with the ether flowing from the syringe held in the right hand.

This liquid evaporates almost instantaneously, and carries along with it the caloric of the water, which is seen, in a little time, to assume consistence, and to pass into a state of congelation; the copper or iron wire is then withdrawn, and brings out with it a small cylinder of ice.

Franklin had made some very curious observations on evaporation, and the effects which it produces, and of the useful application which might be made of it for medical, and even economical, purposes; he has written some very interesting things upon this subject. The experiment of Cavallo is merely an application of this principle; but it is simple, and well adapted for physical demonstrations.

This experiment introduced a conversation on ether, in which I asked this learned philosopher a question relative to an article in Macquer's *Dictionaire de Chimie*, which gave occasion to several persons to attack that celebrated chemist, on the subject of dissolving elastic gum (or caoutchouc) in ether.

"It is certain," said I, "that vitriolic ether, as it is usually prepared, does not dissolve elastic gum. But on the death of Macquer, of whose chemical cabinet I became the purchaser, I found three small decanters, in one of which there was elastic gum, perfectly dissolved in ether; of this, it would be easy to convince any one. The other two contained some, likewise, which appeared to be partly dissolved; but it was precipitated to the bottom in a state a little thicker than turpentine, and was found incapable of mixing with the ether in the bottle. That which contained the elastic gum in a state of perfect solution, had a label with this inscription, in the handwriting of Macquer: '*Elastic gum, dissolved in ether, sent from London.*' I mention this, in order to learn whether you know any one in London who has employed ether in dissolving caoutchouc, and what were the ingredients used in addition to it, or the preparation which it received."

"You could not have addressed yourself," replied Cavallo, "to one able to procure a more complete answer to your questions than myself. We shall go, this morning, to visit the workshops of some artists; and as the person who discovered the process for dissolving elastic gum is in our way, we shall give him a call; your wishes will thus be very soon gratified."

I accepted his offer, and in about an hour after we went to the house of Mr Winch, an apothecary, who received us with a great deal of politeness, and who told me that he was the person who had addressed to Macquer, at Paris, a bottle of elastic gum, well dissolved in ether, and that in a letter to the French chemist, he assured him that the ether did not contain the smallest mixture. Macquer, who found the elastic gum in perfect union with ether, of which the transparency was not in the least altered, and who, on examining the ether, found it to be totally free from any extraneous substance, sincerely believed that pure ether was the real solvent; and notwithstanding his succeeding but imperfectly himself, though he employed the best ether, he probably persuaded himself that what he used was still insufficiently rectified.

"I did not, indeed," said Mr Winch, "send him an account of the process which I used; but it is nevertheless true, that the ether is unmixed, and that the whole depends on a very simple preparation."

Cavallo, who is the friend of Mr Winch, said, that he intended to perform the experiment the next day, at his own house, and that I should be a witness of it. It consists in the following process:—A pound of good vitriolic ether is taken, and put into a bottle, capable of containing about four pounds of any common liquid. On this ether there are poured two pounds of pure water; the bottle is then stopped, held with the mouth downwards, and strongly shaken, in order to mix the two liquids. On discontinuing the shaking, the ether soon swims uppermost; the bottle is still held in the same position, and cautiously opened, keeping the thumb on the mouth of it. The water is by this means easily let off, and collected in a vessel below. The same operation is repeated two or three times, with new quantities of water, until the sixteen ounces of ether are reduced to about five ounces. It is this purified remainder that is found to be the most perfect solvent of elastic gum, which is thrown into the ether, after being cut into small pieces. It begins to swell in a very short time; the ether penetrates it, and appears to act slowly on it at first; but at the end of five hours, or later, the liquid is saturated, and remains transparent. If there be a surplus of elastic gum, it subsides to the bottom; and, on being taken out of the bottle, may be moulded into any form, and will preserve its elasticity.

To shew how the part which is completely dissolved is to be applied to use, I shall describe the method employed by Cavallo to form a tube of elastic gum.

There is first prepared a small cylinder of pipe clay, of the diameter and length of the intended tube. It is not necessary to bake it, but simply to let it dry.

The ether, saturated with gum, is poured into a case of glass, or tin, which should be a little longer than the clay cylinder; this is filled up to the brim.

The operator then plunges the whole length of the clay pipe into the ether, withdraws it suddenly, lets it remain for an instant in the air, replunges it anew, and repeats the operation in proportion to the intended thickness of the tube; for each immersion and evaporation produces a small coating.

This being done, the clay cylinder, covered

with elastic gum, is plunged into a vessel of water; the mould of clay is there speedily dissolved, and the gum remains in the state of a perfect tube.

This method of dissolving and using elastic gum is ingenious; and in one respect resembles that employed by the natives of America, who form all their works in elastic gum, on moulds of clay. It may be objected, that the process with ether is too expensive. The objection holds good with respect to ordinary purposes; but the elastic gum has been so usefully employed in surgery and some other arts, that there are circumstances, in which expense ought to be of no consideration. The process, also, for making ether is so simplified, that it is not half so dear as it was formerly.*

* I should, doubtless, prefer, that we endeavoured to naturalize in Europe the useful tree, or rather trees (for it appears that there are several kinds of them), which yield the singular production of elastic gum. It ought incessantly to be recommended to the naturalists employed by government on voyages of discovery, to direct their attention principally to sending home plants and trees of well-ascertained utility. It is wonderful, that the cinchona (Peruvian-bark-tree), that admirable specific for a multitude of diseases, should still be confined to parts of Peru, where the temperature is not very different from that of many places in the south of France. I must do justice to the administration of the Botanical Garden of Paris; they have not neglected

I should not forget to mention, that the water used in purifying the ether ought to be preserved, because a part of the ether mixed with it may be recovered by distillation.

I saw Mr Cavallo frequently; and it is impossible to be in his company without reaping instruction. He was pleased to make me a present of one of the electrometers which he has brought to such a degree of perfection, and which were not then to be found at the shops of the philosophical instrument-makers. He likewise gave me, with the same obligingness, a beautiful crystal of adamantine spar, which he had from Doctor Lind.* I take the opportunity to repeat my grateful acknowledgments of his kindness.

this object. They have intelligent botanists in America and other places, who have sent home very useful articles; and the economical branch of botany begins to be attentively cultivated by them. It ought not to be omitted, that this garden furnished the first coffee-plant to America. Two plants of that shrub were sent to Declieu, who deprived himself, during a long passage, of a portion of his allowance of water, to preserve the only one which at last remained to him. It arrived, owing to his attention, in good condition at Martinique, where it produced that immense progeny which has supplied all the Antilles. The nation has a debt of gratitude to pay to the memory of this useful man.

* I gave this crystal to the Cabinet of Natural History at Paris.

Doctor Lettsom *

This celebrated physician has a collection of birds, insects, and minerals, some of which are very fine; but of all the objects that are to be seen and admired at his house, the most interesting is, without contradiction, himself.

This friend of humanity, this virtuous quaker, was the first to give the example of emancipating the negroes from slavery, in his rich possessions in America.

He finds the most delightful recompence for this act of justice in the feelings of his own heart, and in the tender and filial attachment of those whose chains he has broken. They have become more inseparable from him since they have had the liberty of leaving him when they please. Happy is the man who places his felicity in doing good to others! We love to meet with such men. They console us for the injustice, and the hardness of so large a portion of our species.

* [Born in the West Indies in 1744; studied medicine in England; completed his training in the University of Edinburgh and in several universities abroad, and took his degree of M.D. at Leyden in 1769. He at once went into practice in London, and soon rose to a high place in his profession, making a large income, which he expended munificently. After a laborious and distinguished career he died in London in 1815.]

All the family of Doctor Lettsom share his amiableness and candour; his circle of friends is of the same description.

After employing a part of the day in administering comfort to his numerous patients, he returns home to share in the enjoyments of friendship, and assembles around him persons whom he loves, and by whom he is beloved.

I supped one evening with him, when some of the most amiable women of London were of the party. It is true, they were neither powdered nor perfumed, and had not, like most ladies, heads full of feathers, or artificial flowers; but their beautiful hair floated with becoming gracefulness on handkerchiefs uncommonly white and fine. The chief attraction of their simple, but elegant dress, lay in the beauty and excellent quality of the stuffs which composed it, and above all in the charming faces and the grace of those who wore it.

Every thing in this house corresponded with that neatness and exquisite simplicity which characterize the quakers. A young widow, of an elegant person, and highly-educated mind, who cultivated poetry, was one of the company. Her agreeable vivacity formed a striking constrast with the mild and tranquil sensibility of the other ladies, all of whom, however, possessed information and talents.

We supped without napkins, a circumstance which is not uncommon in many houses in England; but the best kinds of beer, plain, though exquisitely flavoured, meats, and choice vegetables, were served up in dishes, of an elegant form. The cloth was removed at the desert, and fruits, comfits, and other delicacies, with a variety of wines, in crystal decanters, were placed on a table of the finest mahogany. This is the luxury of the English. We drank more than once in champagne and claret, to the health of our fair companions, and they pledged us in madeira and constance. A lively, but decorous gaiety, mingled with kind attention and the frankest good nature, animated this scene.

Tea, punch, and fine liqeurs, came in their turn. We should have passed the whole night at table, had we yielded to the pressing invitations of the doctor. But, notwithstanding his solicitations, we left the party at one o'clock in the morning. During the remainder of the night, I meditated how I should become a quaker; for, if happiness exists anywhere on earth, it certainly dwells among these worthy people.

John Sheldon *

There are good physicians in Paris, but there are more in London who excel in the practice of medicine. The useful art of curing, or correcting the diseases and infirmities which afflict mankind, requires a preliminary education, so long and so laborious, and depends so much upon the most profound kind of knowledge, in one who seeks to fill this honourable function with distinction, that we cannot too highly appreciate an able medical man.

The English, more wealthy than the French, and consequently much oftener ill, especially in London, where neither the food nor the climate are so salubrious as in Paris, have more need of physicians, and can better afford to reward them for their services. This

* [John Sheldon (1752-1808), a noted anatomist of his day. He went to Greenland to study the anatomy of whales; took a keen interest in aeronautics, and is said to have been the first Englishman to go up in a balloon.]

profession is the most highly respected, and at the same time the most lucrative.

The surgery of the French school is the first in Europe; by the word school I mean the admirable manner in which the French surgeons explain and practise all the branches of their profession; in which they so trace and apply the innumerable ramifications of anatomy, as to enable them to proceed with confidence in the art of curing, and in performing operations which are often terrible to the feelings, but almost always certain and successful in their effects. Thus pain is relieved by pain, and instruments of death are employed in giving life to man.

But in what I have said above, I am far from intending to insinuate that there are not justly-celebrated surgeons in London, and other parts of the three kingdoms. I shall have occasion to mention several of them. I was particularly desirous of seeing those who had made comparative anatomy their study, a subject so intimately connected with natural history.

I had to regret the absence of Doctor Hunter,* who was at this time in the country,

* [John Hunter (1728-1793), the most illustrious surgeon Great Britain has produced.]

but I often visited John Sheldon, and some other anatomists of merit. Mr Sheldon has one of the finest anatomical cabinets in existence. He is known by some excellent publications, particularly a work on the lymphatic vessels, ornamented with magnificent engravings.

This learned anatomist, animated with the passion of prosecuting still farther his enquiries respecting the lymphatics, resolved to encounter the fatigues, and to brave the dangers, of the whale fishery, in order that he might have the opportunity of dissecting the very apparent vessels of these enormous animals.

One must have seen and been well acquainted with John Sheldon, to be able to appreciate his extraordinary passion for study, and the activity of his mind, unceasingly animated by the vivacity, I had almost said, the fervour of his character. There was none of the English gravity about him. I love to meet with such exceptions; but I know that he who is by lively and vigorous conceptions inspired to great undertakings, who labours with ardour, and unites a variety of information to an aptitude and passionate desire for knowing much, cannot have the same uniformity of character, nor act in the same systematic manner with common men.

Sheldon, whom I saw frequently, interested me the more, as he joined to a vivacity, which persons of a colder character might think extravagant, the most estimable qualities.

The discovery of air balloons, excited his enthusiasm. He no sooner learned what had been done at Paris on this subject, than he suspended a part of his anatomical labours, to make calculations respecting the weight of the atmosphere. He afterwards directed his enquiries to the discovery of the most proper substance for making the covering of balloons, to improving the varnish, and to the inventing of the most convenient apparatus for simplifying and perfecting these machines. He visited all the shops and manufactories of London, to gain information on these subjects. He told me that he intended to go to France soon, in order to pay his respects to Montgolfier, Pilatre and Charles; and to see the improvements they had made in the art of aerostation.

But his active mind did not permit him to wait so long before he carried his favourite design into execution, and, in concert with Major Gardiner, he constructed in Lord Foley's garden an aerostatic globe fifty-six feet in diameter made of varnished linen. It was filled with air, rarified by fire. He informed me that he meant this merely as a trial, upon a small scale, calculated to enable him to study this machine; but he was of opinion that experiments would be more satisfactory if they were made, as he hoped they one day would be, upon very large aerostatic globes.

The anatomical cabinet of Sheldon contains a great variety of curious preparations. I spent several mornings in visiting it, and in examining a number of valuable drawings made by able artists; but nothing in this collection interested me so much as a kind of mummy, which was very remarkable in two respects: first, on account of the subject itself, to which I shall immediately refer; secondly, with relation to the manner of the preparation, and the particular care with which it had been made. It occupied a distinguished place in the chamber where this anatomist usually slept; and he was particularly fond of this work.

I was introduced into a very handsome bed-room; a mahogany table of an oblong form, stood in the midst of it, facing the bed.

The top of the table opened by a groove, and under a glass-frame I saw the body of a young woman, of nineteen or twenty, entirely naked. She had fine brown hair, and lay extended as on a bed.

The glass was lifted up, and Sheldon made me admire the flexibility of the arms, a kind of elasticity in the bosom, and even in the cheeks, and the perfect preservation of the other parts of the body. Even the skin partly retained its colour, though exposed to the air.

It appeared to me, however, that the fleshy parts were rather dry, and that there was too great a tenseness of the muscles. This gave to the figure, though it still possessed the remains of beauty, a meagre and feeble air, which considerably diminished the delicacy of its features.

Sheldon informed me that this was partly occasioned by the long sickness of which the young woman died.

He explained to me the manner in which this preparation had been made. He injected the body at intervals with strong spirits of wine, saturated with camphor, and mixed with a small quantity of turpentine.

The skin was prepared and tanned, as it were, with finely powdered alum, rubbed on with the hand. The intestines were taken out, plunged into spirits of wine, and covered with a varnish, composed of a mixture of camphor, and common rosin. The same thing was done to all the internal parts of the body, which were afterwards passed over with alum.

Sheldon assured me, that pounded camphor, mixed with rosin, formed an excellent composition for preserving the flesh, and other soft parts. After having placed all the viscera thus prepared in the body, he then injected the crural artery with a strong solution of camphor, in rectified spirit.

Wishing afterwards to imitate the natural tint of the skin of the face, a coloured injection was introduced through the carotid artery.

In this state of things the body was placed in the table of which I have spoken; but within a double case of timber. The first is made of Virginia cedar (*Juniperus Virginiana*). The inner bottom was covered with calcined chalk, to the thickness of one inch, in order to absorb all humidity. Upon this bed the body was extended. The box, or case, was then carefully shut up, to secure it from access of the outer air.

The box was not opened until five years after the preparation was made. It was then observed to be in the same state of preservation in which it was first enclosed. No mark of decay appeared, and no insect had introduced itself near the body. The box had been several times opened when I saw it; and though this mummy at that time still possessed suppleness in several parts, it is to be supposed that the action of the air will at last completely wither it.

A sentiment of curiosity made me ask Sheldon, at the moment when he was closing up the table, who this young woman was, whose remains he had preserved with so much care. He replied frankly, and without any hesitation, "It is a mistress whom I tenderly loved. I paid every attention to her during a long sickness, and a short time before her death, she requested that I should make a mummy of her body, and keep her beside me.—I have kept my word to her."

I was glad that Sheldon had not informed me of this circumstance sooner, for I confess I could not have avoided experiencing a disagreeable feeling at seeing a lover coolly make an anatomical demonstration on the object of his most tender affection; on a charming young woman whom he had lost, and whose disfigured image could only excite in him the most painful recollections.

I can conceive that there might be a pleasing consolation, a sort of veneration and religious respect, which would extend itself beyond the limits of life, if we were, as in ancient Egypt, in the practice of preserving the remains of our relations, friends, and all those who are most dear to us; but who would, with his own hand, perform the disgusting operations which must be necessary to preserve the body of his friend? I avow I should almost be tempted to act like the Egyptians, who stoned those who engaged in this melancholy business.

But the learned Sheldon does not merit so severe a treatment. He is gentle and compassionate; and I certainly deceived myself, and was wrong in regarding this kind of courage on his part, as an act of cynicism: besides, well-informed persons in London, who were acquainted with this transaction, assured me that it required great strength of mind in Sheldon, to overcome his sensibility. Let us quit, however, this dismal subject, and proceed to describe the dinner which I had with some of the members of the Royal Society.

Dinner at the Academic Club *

About forty members of the Royal Society have been, for more than twenty-five years, in the habit of dining together sociably in one of the taverns of London. Each member has the right of bringing two guests, whom he chooses, among foreigners, or friends of his own acquaintance in the Royal

* [The Royal Society Club, which still flourishes. Formerly, when the Society held its meetings in the evening, the Fellows who were members of the club dined together and adjourned from the dinner to the meeting. At present the Society meets in the afternoon, and the dinners follow after the meetings. See note on p. 49.]

Society. The president may bring a greater number, and can select whoever he pleases for guests.

We sat down to table about five o'clock. Sir Joseph Banks presided, and filled the place of honour. No napkins were laid before us; indeed there were none used; the dinner was truly in the English style.

A member of the club, who is a clergyman (I believe it was the astronomer Maskelyne), made a short prayer, and blessed the company and the food. The dishes were of the solid kind, such as roast beef, boiled beef and mutton prepared in various ways, with abundance of potatoes and other vegetables, which each person seasoned as he pleased with the different sauces which were placed on the table in bottles of various shapes.

The beef-steaks and the roast beef were at first drenched with copious bumpers of strong beer, called porter, drunk out of cylindrical pewter pots, which are much preferred to glasses, because one can swallow a whole pint at a draught.

This prelude being finished, the cloth was removed, and a handsome and well-polished table was covered, as if it were by magic, with a number of fine crystal decanters, filled with the best port, madeira, and claret; this last is the wine of Bourdeaux. Several glasses, as brilliant in lustre as fine in shape, were distributed to each person, and the libations began on a grand scale, in the midst of different kinds of cheese, which, rolling in mahogany boxes from one end of the table to the other, provoked the thirst of the drinkers.

To give more liveliness to the scene, the president proposed the health of the prince of Wales: this was his birth-day. We then drank to the Elector Palatine, who was that day to be admitted into the Royal Society. The same compliment was next paid to us foreigners, of whom there were five present.

The members of the club afterwards saluted each other, one by one, with a glass of wine. According to this custom, one must drink as many times as there are guests, for it would be thought a want of politeness in England to drink to the health of more persons than one at a time.

A few bottles of champagne completed the enlivenment of every one. Tea came next, together with bread and butter, and all the usual accompaniments: coffee followed, humbly yielding precedence to the tea, though it be the better of the two. In France, we commonly drink only one cup of good coffee after dinner; in England they drink five or six of the most detestable kind.

Brandy, rum, and some other strong liqueurs, closed this philosophic banquet, which terminated at half past seven, as we had to be at a meeting of the Royal Society summoned for eight o'clock. Before we left however, the names of all the guests were written on a large sheet of paper, and each of us paid seven livres four sols French money: this was not dear.*

I repaired to the Society along with Messrs Banks, Cavendish,† Maskelyne,‡

* [The Royal Society Club was formally inaugurated on 27th October 1743 under the name of the Royal Philosophers, and the first of its rules was to the effect that dinner should be ordered "every Thursday for six at one shilling and sixpence per head for eating." It was also stipulated that "a pint of wine be paid for by every one that comes." From the text it would seem that the guests paid for their dinners and their share of the drink—an inhospitable practice which has been long obsolete.]

† [Henry Cavendish (1731-1810), grandson of the second Duke of Devonshire; greatly distinguished for his discoveries as a natural philosopher.]

‡ [Nevil Maskelyne (1732-1811); early displayed great powers as an astronomer; took the degree of D.D. and

Aubert,* and Sir Henry Englefield;† we were all pretty much enlivened, but our gaiety was decorous.‡

Doubtless, I should not wish to partake of similar dinners, if they were to be followed by settling the interests of a great nation, or discussing the best form of a government; that would neither be wise nor prudent. But to meet, in order to celebrate the admission of an elector palatine, who has, besides, much merit, to a learned society, is not a circumstance from which any inconvenience can result.§

entered the Church; was appointed Astronomer Royal (1765), and distinguished his tenure of that office by the number and importance of his contributions to science.]

* [Alexander Aubert (1730-1805), Director and Governor of London Assurance Society; fond of astronomy; built a private observatory; became F.R.S. in 1772.]

† [Sir Henry C. Englefield (1752-1822), antiquary and man of science; promoted the publication of admirably illustrated volumes of antiquities, scenery, and geology.]

‡ [At this dinner of the Royal Society Club the author was accompanied by his two future travelling companions, Count Andreani and Mr Thornton. See *Sketch of the Rise and Progress of the Royal Society Club*, London, 4to, 1860, p. 40.]

§ The great Corneille, Moliere, Despréaux, La Fontaine, and Racine, used to take a bottle now and then in a tavern; and they were neither the worse friends, nor the worse poets, for it. How much is it to be wished that some men, who have had sufficient influence in France to destroy the academies by loading them with calumnies, and power enough to re-establish them by bestowing on them their praises, instead of

The Royal Society

The hall in which the meetings of this society are held, is in the old Somerset-house: * it appeared to me much too small. It is not long since it was restored; but notwithstanding the freshness and elegance of the decoration, the room wants that noble and severe character which ought to distinguish a place consecrated to the sciences: it resembles a concert-hall rather than a Lyceum; and the manner in which the seats are disposed, tends to increase this resemblance.

The seats are only simple benches, with backs, ranged in parallel lines, and occupying

flying and abandoning, in the times of misfortune, their unhappy brethren, had sought to assemble them in convivial, but modest, banquets, where their union might have been intimately cemented, and where they might have sworn to defend, with courage and with the arm of genius, the sacred rights of justice and violated humanity: then afflicted France, and indignant Europe, would not have had to regret those illustrious and unfortunate victims, who have been delivered up to ferocious tigers; and we should still have counted, among the learned men who honour their country, Malesherbes, Bailly, Lavoisier, Condorcet, and many other philosophers and artists, who have been cruelly butchered.

* [The Royal Society removed from its old home in Crane Court, Fleet Street, to Somerset House in November 1780, and remained there until 1857, when it was transferred to Burlington House, Piccadilly, where it now occupies a suite of apartments specially built for its use.]

the whole of the room. The president and the secretaries have alone distinguished places. The former is seated in an elevated chair, of a colossal form. It is made of mahogany, and surmounted with an escutcheon, on which are painted the insignia of the society. Nothing can be in a more gothic or worse taste than this ornament.*

The table, which is before the president's chair, is elevated, and covered, one cannot tell why, with an enormous cushion of crimson velvet. Before this, there is a second table, destined for the secretaries, upon which there lies a large mace of gilded wood, or metal.† This is the symbol of all the royal institutions.

The meeting began precisely at eight o'clock. Sir Joseph Banks presided, and Blagden ‡ was one of the secretaries. The strangers were placed near the members who

* [The arm-chair thus disparaged is still the time-honoured presidential seat at the meetings of the Society. It appears as a background in several portraits of past presidents suspended in the Society's rooms, among them, Sir Joseph Banks.]

† [This mace is of silver, richly gilt, and weighs 190 ounces avoirdupois. It was presented to the Royal Society in 1663.]

‡ [Sir Charles Blagden (1748-1820); studied medicine at Edinburgh and became an army medical officer. Elected F.R.S. in 1772, and became Secretary in 1784.]

introduced them; and, however little known they might be, every member behaved to them with the greatest politeness and affability.*

The president first read the names of the strangers admitted that night, and the names of the members who had presented them. He afterwards proposed a ballot for the Elector Palatine to fill a vacant place. As the result of the voting the elector was admitted with applause. When this business was finished, the meeting broke up.

Some members whom I had the honour of knowing, engaged me to go next day to Greenwich, to see the Observatory, where a committee of the society was to meet, by the order of the government, to examine the state of the astronomical instruments. It is the practice to form committees of this kind annually; for in this country whatever is connected with naval affairs, becomes an

* The detractors of the English character, have inappropriately reproached them with behaving in a cold and surly manner to foreigners. What has been considered as coldness, is only reserve. Strangers were politely and honourably received in this learned society, and placed by the side of the members with whom they were fraternally mingled. The sciences, like the muses, should be sisters, and ought to know no distinction of country or of government.

object of general attention, and is never lost sight of for a moment.

After this visit, there was to be a dinner in the country; and Herschel,* who was one of the committee, was expected to be there. I was to be introduced to this illustrious astronomer, and I had some hopes that he would not refuse me the favour of seeing the large telescopes in his observatory in Windsor Forest.

The Observatory of Greenwich

This useful institution, which is consecrated to astronomical observations, is situated on a hill, about seven miles from London. I went there in a coach, which arrived in about an hour and a half.

The Observatory is built on the most elevated part of the hill; and it affords one of the finest views imaginable.

The Thames at your feet, covered on

* [Sir William Herschel (1738-1822), born at Hanover; came to England in 1757; acted first as church organist, then turned his attention to the construction of telescopes, which he himself used in observing the stars. His fame as an astronomer rapidly grew as his remarkable discoveries succeeded each other, and was shared with him by his sister Caroline Lucretia (1750-1848), who not only aided him in his researches, but distinguished herself by original astronomical observations of her own.]

either side and all its length with a triple row of vessels—streamers of various colours waving in the air—ships under sail, going up and down the river—an immense multitude of men of all nations on this floating city—the masts in the distance mingling with the steeples—three high bridges, one beyond the other—the church of St Paul, whose dome and fine proportions excite admiration even at this distance—Westminster Abbey, with its towers and Gothic architecture—the column, called the Monument, rising from a nearer level to the height of two hundred and two feet—all these grand and magnificent objects form a picture for which the true point of view is at the Observatory of Greenwich.

The building is made of brick, in a style of the greatest simplicity. Magnificence and research are displayed in the inside only in the perfection of the instruments, which nothing can exceed.

I found the committee assembled, and Doctor Maskelyne, the director of the Observatory, had the goodness to shew me, in great detail, the most remarkable objects in that rich collection.

The truly learned are easily distinguished by their manners; nothing can equal their compliance and affability: this rule is hardly ever found at fault, for the cultivation of the mind softens the manners, as that of the earth renders its productions more delicate. The man of science, or of literature, who should appear proud, haughty, or self-sufficient, must always seem to the man of sense a mediocre being, if not mentally diseased. Great timidity, habits of retirement, and the importunities to which celebrated men are exposed, may, indeed, be sometimes taken for coldness; but the difference is easily distinguishable.

Dr Maskelyne added to his other kindnesses that of introducing me, along with Mr Banks and Mr Aubert, to Mr William Herschel, who was so good as to invite me to see his observatory, and the large telescopes of his construction, at his country-house, and arranged a day for the visit.

At four o'clock, the committee having finished its business, we all assembled in a famous tavern, in the neighbourhood. There were about thirty persons at table. The dinner was served in the English fashion, and seasoned, according to custom, with a preliminary benediction on the guests and the victuals. The repast was excellent, and especially gay and extremely agreeable.

We rose from the table at seven o'clock, not to depart, but to pass into another room, where cut bread and butter, tea, coffee, brandy and rum awaited us, disposed with much show upon a large table. The tea is always excellent in England; but nowhere do people drink worse coffee. The English must be little sensible of the delicious flavour of this agreeable beverage, which nature seems to have created to solace at once the body and the mind: it not only strengthens without injuring the stomach, but reinvigorates without enfeebling the spirit. Voltaire, who was extremely fond of coffee, called it with good reason the *quintessence of the mind.*

Why then does the English government, for political and commercial reasons, prevent the people from using coffee which they might prepare according to their own taste, and compel them to purchase from monopolist dealers a kind of inferior quality, and bad flavour, which has been roasted a long time before. One would imagine that it has been purposely contrived to make this beverage disgusting which is so capable of removing melancholy in a country where the atmosphere is wrapped in an almost funereal gloom—and where, if we may believe Fielding, there is more port wine drunk in one year, than Portugal produces in three.

It would certainly be far better policy to substitute for tea, which must be brought from China, the coffee which grows in the English colonies: such a change might, perhaps, tend to diminish that alarming consumption of wine which occasions in this country so many diseases and especially so many excesses caused by drunkenness.

I beg pardon of the reader for this digression, which is somewhat foreign to my subject: but I was so disgusted with the bad coffee which I found, even in the most opulent houses of London, that, on account of my attachment to the better kind of that beverage, I could not avoid paying it this small tribute of gratitude, or, if the word should be preferred, of epicurism.

CHAPTER II

Sir Joseph Banks' Country-house. — The Observatory of William Herschel, near Windsor.—His large Telescopes. — Miss Caroline Herschel, his Sister

ON the 15th of August, which was a fine day, I made an excursion to the country-house of Sir Joseph Banks, about ten miles from London, where I saw his gardens, several objects of cultivation and management which interested me, and likewise a beautiful bird, which had never before been brought alive to England. This bird was the green pigeon of the island of Nicobar: its plumage, of a deep green colour, shines with a brilliant lustre, and its liveliness of disposition is as striking as its colour. It is bold, petulant, and has none of the gentle manners characteristic of the dove family. Although an exception to the general rule, this fine bird is all the more curious. It is of the size of a large ordinary pigeon, but with a longer body. It is said to be excellent eating.*

Sir Joseph Banks informed me, that the seamen who had brought away some of these birds from the island of Nicobar, intending to sell them in England, could not resist the temptation of eating them during their voyage. This one had accidentally escaped from being devoured, and was the only one saved out of a considerable number.

It were to be wished that they had left him a companion: this superb species would, perhaps, have been propagated in Europe; at least, the experiment might have been made.

At seven in the evening, after an elegant dinner and a dessert, at which there was abundance of pine-apples, I took leave of Sir Joseph, and set off to meet William Herschel, who expected me. Count Andreani and William Thornton were of the party. The country-house in which Mr Herschel

* *Columba Nicobaria*, of Linnæus, *Syst. Nat.* page 283, 27. —*Columba Nincombar Indica*, of Klein, page 120, 28.—*Pigeon of Nincombar*, of Albin, vol. iii, page 20, with the figure of the male; plate 47, the figure of the female; both badly coloured. —*Pigeon of Nincombar*, of Brisson, vol. i. p. 153, 44, no figures.—*Pigeon of Nincombar*, of E dwards; *History of Birds*, plate 339, a pretty good figure.

makes his observations stands at one end of the forest of Windsor, about twenty miles from the house of Sir Joseph Banks; but, with good horses, and in an English carriage, the journey may be performed in three hours.

It was about the time when highwaymen usually come upon the road, to rob the imprudent traveller. They are known to be numerous, and to carry on their dangerous profession upon horseback, often even mounted on hunters; but we were informed that, though our danger would have been great on the evening before, we were safe that night, which was Sunday, when the roads are covered with people of all ranks, who, having passed the day in the country, return at night to London, to be ready to resume their usual occupations on Monday morning.

The evening was most beautiful,—the weather calm and mild, the sky sparkling with stars. The road, as carefully made and as smooth, as the avenue of a public promenade, was bordered with hedges, almost all in flower, and serving to inclose charming gardens and parks, ornamented with exotic trees, in the midst of which many simple, but delightful houses seemed to dispute the ground with each other.

The road was, at this time, covered with numerous cavalcades of men and women, with many servants in their train. Carriages of every kind, most of them very elegant, but all of them substantial and commodious, and many of them, with superb equipages, succeeded each other without interruption, and with such rapidity, that the whole picture looked like magic: it certainly showed a degree of wealth and extent of population, of which one has no notion in France. All was life, movement, and rapidity; and, by a contrast only to been seen here, all was calm, silent, and orderly. A tacit and inviolable respect for each other reigned among the individuals composing this impetuous whirlwind of people all hastening towards the same point. The kind of religious silence pervading this extraordinary scene, faintly illuminated by the mysterious glimmer of the stars, transports one who sees it for the first time into the happy fields of Elysium.

But the story of Elysium is fabulous, and that which I relate here is real; for it is what I have seen and proved, what all

Englishmen, and those who know this astonishing country, will acknowledge to be a just description. Whence then comes this calm in the midst of so much movement? It has its origin in the state of the public mind, which is well formed; in the education of the nation, which is good; in reflection always active; in the forms of worship, which, stripped of all vain superstition, consecrates the day of rest to pious meditation, while the law comes in aid by severely repressing the noisy games and tumultuous orgies which on this day degrade or brutalise men in nearly all catholic countries.

Country-house of William Herschel.—His Sister.—His Telescopes.—Observations at Night

I arrived about ten o'clock in the evening at the door of this celebrated astronomer. I entered, by a well-lighted staircase, into a room adorned with maps, instruments of astronomy and natural philosophy, spheres, celestial globes, and a large harpsichord.

Instead of the master of the house, I observed, in a window at the farther end of the room, a young lady seated at a table, which was surrounded with several lights; she had a large book open before her, a pen in her hand, and directed her attention alternately to the hands of a pendulum-clock, and another dial placed beside her, the use of which I did not know: she then noted down her observations.

I approached softly on tiptoe, that I might not disturb a labour, which seemed to engage all the attention of her who was engaged in it; and, having got gradually close behind her without being observed, I found that the book she consulted was the Astronomical Atlas of Flamsteed, and that, after looking at the dials of both instruments, she marked, upon a large manuscript chart, points which appeared to me to indicate stars.

This absorbing employment, the hour of the night, the youth of the fair student, and the profound silence which prevailed, interested me greatly. At last she turned round her head accidently, and discovered how much I was afraid to disturb her; she rose suddenly, and told me how sorry she was that I had not informed her of my arrival, that she was engaged in following and recording the observations of her brother, who expected me, and who, in order that he might not lose the precious hours of so fine a night, was then busy upon the stars in his observatory.

"My brother," said Miss Caroline Herschel, "has been at work these two hours; I do all I can to assist him here. That clock gives me the time, and this dial, the hand of which communicates by strings with his telescopes, informs me, by signs which we have agreed upon, of whatever he observes. I mark upon that large chart the stars which he enumerates, or discovers in particular constellations, and even in the most distant parts of the sky."

This fraternal intercommunication, applied to a sublime but abstruse science, this activity, this persistence during successive nights in great and difficult observations, afford a pleasing and impressive example, calculated to excite an enthusiasm for the sciences, since they present themselves under an aspect so amiable and so interesting.

Mr Herschel's observatory, to which I repaired some moments after, is not built on an eminence, nor on a high building; he has preferred to place it on a carpet of verdure, where no movement can shake his instruments, and which is sufficiently extensive to permit all the motions such large machines require.

His telescopes are elevated in the air, and mounted in a most simple and ingenious manner: a young man is placed in a kind of chamber below, who, by the means of machinery, turns the telescope and the observer together in a circle, with a gradual motion corresponding to that of the earth; thus the reflexion of the star observed is retained in the field of the metallic mirror.

These large pieces of mechanism are, besides, constructed with so much precision, solidity, and care, that they can withstand the inclemencies of the air, as well as hoar-frost; and the mirrors are so disposed, that they can be taken out and replaced at pleasure, with the greatest facility, notwithstanding that they are of considerable weight.

Here I saw that ever-memorable telescope with which the eighth planet was discovered.* Mr Herschel gave to it the

* It was discovered in 1781; its motion is from west to east, like that of the other planets. By observing it attentively with the largest telescopes, Herschel has discovered two satellites moving round the planet, in orbits nearly circular, and almost perpendicular, to the plane of the ecliptic.

name of the King of Great Britain, and called it *Georgium Sidus*. But all astronomers, actuated by a feeling of general gratitude, have, with one unanimous voice, unbaptised it, if I may use the expression, and have given it the name of *Herschel's planet*.*

This telescope, with which I had the pleasure of making observations during more than two hours, is only seven feet long by six inches six lines in diameter. Mr Herschel assured me, that he had cast and ground more than one hundred and forty mirrors with his own hands, before bringing the instrument to the last degree of perfection. A telescope six feet long is placed beside this one.

This celebrated astronomer has not confined the size of his telescopes to the last-named measure; two others, twenty feet long, and mounted on a large standard, are

* Laplace, in his learned work, *L'Exposition du Système du Monde*, calls this planet *Uranus*. My correspondence with several members of the Royal Society of London having been suspended since the commencement of the war, I am ignorant of the reason of this change; but I persume, that it is owing to the modesty of Mr Herschel, who has, doubtless, refused an homage so justly, and so universally, paid to him by the learned world.

imposing machines which rise higher than the house. The diameter of one of these telescopes is eighteen inches and three-quarters, while its mirror weighs one hundred and fifty pounds.

As these superb instruments are of the Newtonian kind, which require the observer to be beside the object glass, Mr Herschel has constructed an apparatus of ingenious mechanism, by which the farther end of the telescope can be reached with ease and safety. There the observer finds a turning chair so disposed, as to enable him to sit at his ease, and follow the course of the stars. A domestic, placed below the telescope, by means of an ingenious combination, moves it softly and gradually, along with the observer and all the apparatus.

It is thus William Herschel has been enabled to discover distinctly those innumerable stars, which form the most pale and distant part of the Milky Way.

It is thus that he has been enabled to observe that multitude of double stars, as well as so many nebulæ, with respect to which only vague and uncertain notions had previously prevailed. Thus, too, has he undertaken to count the stars of the heavens, and has made such astonishing discoveries.

Placed at the upper end of his telescope, when the indefatigable astronomer discovers in the most desert parts of the sky a nebula, or a star of the least magnitude, invisible to the naked eye, he informs his companion of it, by means of a string which communicates with the room where she works; upon the signal being given, the sister opens the window, and the brother asks her whatever information he wants. Miss Caroline Herschel, after consulting the manuscript tables before her, replies, "Brother, search towards the star *Gamma*, towards *Orion*," or near such or such constellation which she names. She then shuts her window, and returns to her employment.

One must be born with a very great indifference for the sciences, not to be affected by this delightful accord, and not to wish that the same harmony should reign among all those who have the happiness to cultivate them. How much more rapid would their progress then be!

We commenced our observations with the *Milky-way*.

The telescope of twenty feet discovered to us, in the palest and most distant part of the heavens, an immense number of bright stars, quite distinct and separate from each other.

Mr Herschel then directed the instrument to the star in the foot of the Goat, which emits so strong a light as to affect the eye. On making its light fall upon a paper written in very small characters, we could discern and count the lines with ease. It is curious thus to distinguish objects by the glimmering of a star, that is, a sun, distant many hundreds of millions of miles from the confines of our system.

The double stars, which are not visible with the most powerful achromatic glasses, appear separate and quite distinct, when viewed with the telescope of twenty feet long.

Mr Herschel made me observe the nebulæ of M. Messier, at first with the telescope of seven feet, that is, with the one which served to discover the planet. These little specks appear still nebulous with that instrument; and one perceives only a feeble and confused glimmer. But with the telescope of twenty feet one can no longer doubt that they are clusters of stars, which appear confused only from their immense distance; by this telescope they are found to be perfectly distinct.

Mr Herschel requested me to direct my principal attention to the stars which he was the first to discover to be of different colours from each other, and among which some are seen that border on blue, others on orange, and several on a bluish tint, &c.

It is certainly neither to an optical illusion, nor to the effect of the mirrors and lenses, which Mr Herschel uses, that we ought to attribute this difference of colour. I started every possible objection upon the subject; but the learned observer always answered them by facts, to which it was unreasonable to reply. Thus, for example, he repeatedly directed the telescope to two double stars of nearly the same magnitude, and separated from each other by a small interval only; that is, small in appearance, for the interval must be immense if we consider their distance from the earth. Both were of the same colour, and emitted the common white light of the stars.

On directing the same telescope immediately after to other neighbouring double stars, the one was evidently of a blue colour, the other of a silvery colour. The blue star was in some instances on the right, and in others on the left. I saw also some single blue stars, several of a bluish white, and others of an orange colour.

Mr Herschel said to me with much modesty, that this discovery was not of very great merit, since it was so easy to make it without recourse to large telescopes; acromatic ones with large object-glasses being sufficient to show the coloured stars above mentioned.

The observations, however, of Mr Herschel were at first disputed, for it is much easier to deny than to examine. But they were soon confirmed, as they deserved to be, by the astronomers of Germany and Italy, and at the observatory of Paris, by M.M. Cassini, Mechain, &c.

Mr Herschel shewed me a considerable work on the stars; which he designs to publish as soon as it is brought to a conclusion. He has confirmed what had been previously observed, that several stars distinctly indicated in the ancient catalogues, and of which some are even engraved in the celestial Atlas of Flamsteed, have entirely disappeared. It is thus probable, that there sometimes happen great revolutions and perhaps terrible catastrophes in some parts of the system of the universe; since there are suns which become extinct, and thereby plunge into annihilation any organised beings that existed on the planets which these same suns illuminated.*

Jupiter, viewed with the telescope of twenty feet, appears much larger than the full moon.† His parallel belts are very distinct, and his satellites are of a truly astonishing magnitude.

* "Sometimes," says Laplace in his *Exposition du Système du Monde*, tome i. page 88, "stars have been observed to appear on a sudden, and then to disappear, after shining for a while with the strongest lustre. Such was the famous star observed in 1572, in the constellation of Cassiopea. In a short time it surpassed the brightness of the most splendid stars, and even of Jupiter itself. Its brightness then decreased, and sixteen months after its discovery it entirely disappeared, without changing its place in the heavens. Its colour underwent several considerable alterations; having been at first of a bright white, then of a reddish yellow, and last of all of a lead-coloured white."

† This is not astonishing, when we reflect that Jupiter is at least a thousand times bigger than the earth. "Jupiter," says Laplace, "is at least five times farther from us than the sun. When its apparent diameter is 120 seconds, the diameter of the earth would not appear, at the same distance, under more than an angle of 11 seconds. The bulk of Jupiter is therefore, at least, a thousand times greater than that of the earth." *Exposition du Systeme du Monde*, tome i. page 78.

On directing the same telescope towards Saturn, we saw his ring in the most distinct manner, and also the shadow which it projected on the body of that immense and singular planet. Mr Herschel shewed me the sky, and even several stars, in the interval between the moveable ring and the planet. By means of some luminous points which are remarked in the ring, he was enabled to discover that this solid circle has a rotation from west to east like that of all the other planets of our system.

The micrometer which Mr Herschel uses is composed simply of two threads of silk, very fine, well stretched and parallel, which may be moved to a greater or shorter distance at pleasure. This instrument was known before, but this acute observer has perfected it, by finding an easy method of turning one thread over the other at pleasure; so that, on placing them in the telescope, he can take angles with the minutest precision.

The inventor of such large telescopes is far from having confined himself to those of twenty feet long. He was engaged in making the necessary preparations, to construct one of forty feet in length, and of a proportionable diameter.*

Mr Herschel's intention in constructing telescopes of this great size, is not so much to magnify the object, as to obtain, by the aid of mirrors of such a vast field, a more considerable quantity of light. This project is new and excellent. He told me, that he expected to encounter great difficulties in carrying to perfection a telescope of that dimension and weight; but that he, at the same time, expected such great effects from it, that nothing would discourage his progress.†

* The mirror is four feet in diameter, and weighs two thousand weight: the telescope, with its apparatus, weighs forty thousand weight. "It gives so much light," says Lalande, who saw it when finished, "that the nebula of Orion seen through it, emits a brightness like that of noon day." *Astron.* tome ii. p. 635, *nov. edit.*

† Since that period, this astonishing telescope has, with immense labour, been carried to the highest point of perfection. William Herschel himself, in a letter to Mr Watson, dated the 10th of September, 1791, speaks of it as follows:—"I have, as usual, been much occupied in polishing mirrors, for telescopes, of all sizes, in order to bring to perfection that difficult part of optics. It would, truly, be impossible to conceive the time which I have spent, and the pains which I have taken, to accomplish this end. But I have been amply compensated by the pleasure which always accompanies the pursuit of a favourite object, and by the success which I flatter myself I have obtained. My telescope of forty feet is actually the best instrument I have in my possession; that is to say,

I remained until daylight in this astonishing observatory, constantly occupied in travelling in the heavens, with a guide, whose boundless complaisance was never wearied by my ignorance, and the importunity of my questions. I passed about seven hours there, employed without intermission in observing the stars. It was impossible to think the time long, when spent in a manner to me so profitable and so curious. That delightful night appeared no more than a dream to me, and seemed to last only a few instants; but the remembrance of it will be indelible;

it enables me to observe better than any of my other telescopes, those objects which are most difficult to be seen distinctly, such as Saturn, his satellites, and his ring, or rather his rings; for I have lately transmitted to our President a paper relative to that planet, in which I have clearly shewn that it has two distinct rings, separated from each other by a considerable distance—such, that with my telescope of forty feet, I have seen the sky very distinctly through this space, the extent of which is 1741 of our miles: the diameter of the exterior ring, measured with the same instrument, appears to me more than 222 of our miles. I have also proved, in the same paper, that the fifth satellite of Saturn turns round its axis in 79 days, 7 hours, and 47 minutes; a time equal to that of its revolution round the planet: thus its movement, in this respect, exactly resembles that of the moon, whose revolution on her axis is made precisely in the same time she takes to pass round the earth." *Journal of Experimental Philosophy and Natural History*, 1792. It ought not to be omitted, that Mr Herschel has discovered two new satellites belonging to Saturn, with the same telescope.

and my grateful recollection of the kindness with which Mr Herschel, and his interesting sister, condescended to receive me, will never be erased from my heart.

I left Slough (the name of the place where Mr Herschel resides) about eight in the morning, to go to Kew; where Sir Joseph Banks had appointed me to meet him, for the purpose of shewing me the whole of the superb gardens of that place, particularly that of botany.

Gardens of Kew

This charming place is seven miles from London. I shall not enter into a description of the house, or of the pleasure-gardens, the temples, the bridges, the towers, with which it is adorned; I shall confine myself to what relates to the garden for botanical instruction. George III. has been careful to assemble there the rarest vegetable productions from all parts of the globe. He has thereby rendered a signal service to botany, as he has conferred a no less important one on astronomy, by so honourably encouraging the labours of Herschel, and enabling him to construct the largest and most perfect telescopes which mankind has ever yet witnessed.

The gardens of Kew are so well laid out, and so well kept, the order and the taste which pervade them are so admirable, and art has here so studiously endeavoured to resemble all that is beautiful and striking in nature, that I place this garden above every thing of the kind that I have ever seen.

The weather was superb, and the season had been so favourable, that the rarest and most delicate plants displayed a luxuriancy and variety of foliage, blossoms, fruits, and perfumes, which formed an enchanting whole.

What most attracted my admiration in this garden of foreign plants and exotic trees, was an exquisite neatness, a pure taste, and a judicious order, which must there render instruction attractive. The mingling of the trees and shrubs of both hemispheres is so well planned, and so perfectly harmonized, that the mind seems to rest every where with the same satisfaction, the same soothing yet varied sensations; every thing charms, and nothing fatigues it.

The green-houses are disposed with much judgment. Some of them have only a moderate heat, for plants which thrive in a mild temperature; to others a strong dry heat is given, suited to those of the climate of Africa; while those designed for plants found in parts of America, where the atmosphere is loaded with vapours, receive a moist heat. It is with all these precautions, and these imitations of nature, aided by incessant care, that plants the most precious, and the most difficult to preserve, grow here almost as well as in their native soil.

I saw, with considerable interest, in one of the green-houses, a curious plant, which had lately come into flower; it was the *Hedisarum girans*, brought from the East Indies, in 1775, by Dr Patrick Russel.

This tall and elegant plant is endowed by nature with a sensibility so remarkable, that, if placed under glass frames, inaccessible to the air, about mid-day, when the sun is most powerful, its lateral leaves, which are formed in the shape of a spear, exhibit a spontaneous gradual and alternating movement of ascent and descent, such as might be thought to be the effect of art.*

* *Hedisarum foliis ternatis ovali-lanceolatis, obtusis lateralibus imminutis.* Aiton, Hortus Kewensis. *Hedisarum girans. Linn. Supp.* 332.

Another species of the *Hedisarum*, brought from Cochin China by Sir Joseph Banks, was likewise in blossom at the same time. Its leaves are of a form so extraordinary, and of a tint so fantastic, that it has received the name of the "*Bat Hedisarum*"; but the contour of the leaves, their lightness, and their colour, have a much nearer resemblance to the wing of a butterfly.*

Amidst a multitude of rare and singular plants, one of them attracted my particular attention: it was the *Dionea muscipula*. I had seen it once before in the Jardin des Plantes of Paris. Franklin had it sent over in its native state from the marshes of South Carolina as a present to Buffon. It arrived in good condition: but it was so delicate, that it lived only six months. In the garden of Kew, however, the plant was in the best possible state of vegetation.†

* *Hedisarum foliis simplicibus ternatisque; foliis intermediis bilobis; lobis lanceolatis divaricatis, leguminibus plicatis.* Aiton, Hortus Kewensis. *Hedisarum vespertilionis. Linn. Supp.* 331. *Bat-winged Hedisarum, nat. of Cochin China. Introd.* 1780, *by Sir Joseph Banks: fl. July and August.*

† *Dionea muscipula, cal.* 5 *phyllus. Petala* 5, *cap.* 5, *unilocularis, gibba, polysperma.* Aiton, Hortus Kewensis. *Dionea, Linn. Mant.* 238. *Venus' fly-trap; native of Carolina, fl. July and August.*

This extraordinary plant has thick leaves, like those of certain oily plants. They are disposed in the form of hinges, covered with prickles, and furnished by nature with a honied substance. The flies, attracted by the sweetness of this liquid, come to suck it; but the plant is endowed with such acute sensibility, that it is irritated by the smallest touch; the leaf doubles up its folds, shuts upon itself, seizes the insect with its prickles, pierces and kills it. Nature thus appears as inexhaustible in her means of destruction, as in her means of creation.

The *Magnolia grandiflora*, planted in open ground, and forming trees of a great height, were covered over with their beautiful flowers, which perfumed the air. Their foliage of a bright green above, and of a pale yellow and variegated colour below, produced the greatest effect, contrasted with trees of a silver-coloured, and others of a reddish, foliage.

Evergreens, resinous trees of all kinds, loaded with their scaly fruit, and of various shapes, were intermingled with those of a soft green, with parasol-shaped trees, with some of downy, some of indented, and others of palmated, leaves. From these various intermixtures, artfully combined, and tastefully arranged, there resulted a variety of forms, tones, and colours, which produced the most striking contrasts, without ever exhibiting an injudicious contrariety.

The polypodies, ferns, and different plants which require coolness and the shade, are placed in appropriate situations. The heaths, honeysuckles, brooms of various kinds, ivies, and myrtles, appear in their vicinity.

But nothing astonished me so much as the admirable art, by which mosses, the most delicate capillary herbs, and even some lichens, have been raised with such success as to unite in one place the most complete and the best assorted display of the richest vegetable productions of nature.

To accomplish this purpose, there was collected a great quantity of lavas, of which Sir Joseph Banks had brought a plentiful supply on his return from Iceland, where he had gone to visit the volcano of Mount Hecla. The ballast of his vessel consisted entirely of these lavas; and it was this circumstance that suggested the fortunate application that has been made of them. As the lavas are full of cavities, fissures, and roughnesses, and are likewise spongy, and capable of imbibing and long retaining water, it was resolved to form thick borders of them, more or less elevated, round the verges of a shady piece of ground, appropriated to this moss-garden, which is unique of its kind.

This numerous family of *Cryptogams*, so varied in form, in colour, in their mysterious and wonderful fructification, grows up and thrives in the cavities of these little artificial rocks in a manner which appears almost miraculous, and which does honour to the taste and intelligence of him who first conceived the happy idea.

The experiment having perfectly succeeded, they wished to practise it on a greater scale. To effect this, the lavas imported from Iceland being exhausted, they had recourse to factitious lavas, which were formed of lumps of clay, vitrified in a strong coal-fire, and which were found to answer nearly the same purpose.

It must be allowed, that the climate of England, which is very favourable to the growth of those plants, contributes much to the success of this pleasing invention. I am of opinion, however, that similar attempts may be successfully made elsewhere, and that the great botanical gardens may thus be enriched with an assemblage of objects, which cannot, in general, be studied but in herbals, or by performing voyages, which one is not always in circumstances to undertake.

Mr William Aiton, who is Director of this magnificent garden, and who has contributed to bring it to its present state of perfection, is engaged in writing a description of its numerous and rare plants, which he has cultivated with so much knowledge, zeal, and application. This work is expected with impatience, and will be received with interest by botanists, and by all those who know how to estimate the talents of Mr Aiton.

This modest naturalist exerted himself in the most affable manner, as did also Sir Joseph Banks, to point out and explain every thing that could interest my curiosity. I felt the more indebted to their extreme complaisance, as my ignorance obliged me to trouble them repeatedly with tiresome questions; and I gladly seize this opportunity to renew my apologies, and my thankful acknowledgments.*

British Museum

This immense collection of scientific and curious objects is deposited in the vast mansion of the late Duke of Montague in Great Russel-street.

This museum is composed of manuscript and printed books; of Egyptian, Etruscan, Greek and Roman antiquities; Indian, Chinese, and Japanese idols; of the clothing, weapons, and utensils of the islanders of the South Seas and other savage peoples; of quadrupeds, amphibia, birds, insects, fishes, shells, and other marine productions; of minerals, petrifactions, and fossils of every kind.

This immense assemblage of objects was partly formed by the celebrated Hans Sloane. It is a pity that the collection was not al-

* Mr Aiton [1731-1793] published, in 1789, an excellent work, containing a description of the plants of this garden, under the title of *Hortus Kewensis, or, a Catalogue of the Plants cultivated in the Royal Botanic Garden at Kew. By William Aiton, 3 vols. in 8vo fig. London,* 1789. A short time after the publishing of this book, death deprived his friends and botany of this estimable naturalist. His son succeeded him in the management of Kew Gardens.

lowed to remain as he originally left it. Had no additions been made to it, and had it been allowed to retain the modest title of *Sloane's Museum*, there can be no doubt that people would have been anxious to visit the collection of that indefatigable naturalist, and would have viewed with as much astonishment as satisfaction, what can be done by a passionate love of science, aided by an affluent fortune and a noble generosity.

But I am not pleased that the collection of a private individual, to which there has been since superadded a crowd of heterogeneous objects, calculated rather to distract, than to command the attention, should possess the title of *The British Museum*.

A nation, worthy of commendation for the highly advanced state of her commerce and manufactures, and the importance of her navy, the results of a multitude of difficult combinations and profound knowledge, ought to have monuments worthy of herself, and more akin to the greatness and boldness of her character.

The English have been reproached with not giving sufficient encouragement to the sciences, and especially with not investing them with the consideration which they deserve. I cannot, I must confess, decide upon that charge. But if it were well founded, the government would seem very little attentive to that in which it is so deeply interested; for those who sit at the helm of affairs are a thousand times too enlightened to be ignorant that England has reaped more real glory and distinction from the uncommon geniuses she has produced, than from her conquests in the two Indies, from her fleets, her battles, or her eternal parliamentary discussions.

All this political scaffolding, notwithstanding its aim at utility, will have a thousand times disappeared, whilst the name of the immortal Newton,* that of Napier,† of

* The glory of the English nation—an immortal genius, who claims pre-eminence over all others, from his discovery of the principles of the celestial movements, and the laws of universal gravitation. His *Principia* is the greatest and most astonishing work ever produced by the human mind.

† Napier, a Scottish Baron, who was the inventor of Logarithms, an admirable contrivance, "which," says Laplace, very justly, "by reducing the labour of many months to that of a few hours, has in a manner doubled the lives of astronomers, and saved them from the errors and vexation inseparable from long calculations;—an invention so much the more satisfactory for the human mind, that it has derived it wholly from its own self. In the arts, man employs the materials and the powers of nature to increase his own powers; but here every thing is his own work."

Halley,* of Bradley,† and other illustrious philosophers, will be held in veneration by all nations among whom war and the homicidal frenzy of conquest shall not have extinguished the torch of light and knowledge, which leads to truth, the sole object of man in the rapid course of his life.

The British Museum contains many valuable collections in natural history; but with the exception of some fishes in a small apartment, which are begun to be classed, nothing is in order, every thing is out of its place; and this assemblage is rather an immense magazine, in which things seem to have been thrown together at random, than a scientific collection, intended to instruct and honour a great nation.

It may be presumed, that as long as so repulsive a confusion is suffered to continue, no artist will ever be induced to go there, to acquire those branches of information which relate to the materials he uses, and the sources whence they are derived.

* Halley, whose admirable investigation of the Cometary System enabled him to discover and to predict the return of the comet which appeared in 1769.

† An astronomer, for ever celebrated by his discovery of the aberration of the fixed stars, and the nutation of the earth's axis.

Never will the painter repair thither to see and to study animals according to nature, and to admire the different tones of colour, and the infinite variety of shades presented by the plumage of birds, the gay attire of butterflies, and the oriental splendour of shells.

Never will the physician, who devotes his nightly studies to the cure or alleviation of the diseases of his fellow-creatures, go to learn from that chaos the means of distinguishing with precision the beneficent productions, which in the various climates of the globe nature seems to offer to man, in order that he may mitigate or remove the many evils which everywhere accompany him.

The philosopher, who loves to behold nature on a great scale, and he who delights in studying the details of that immense chain which seems to connect together all created beings, and of which the last link appears to join on to the first, will find nothing that can interest him in the midst of such disorder.

In a word, youth, so inquisitive, and so fond of novelty, will here find no excitement to study by the attractive lure that is so capable of cultivating his tastes when these are flattered and captivated by a collection wherein order and neatness reign throughout all its arrangement.

But that which has no existence at present, may one day be accomplished. I sincerely wish it for the progress of natural science; seeing that a nation, whose political and commercial relations extend to both hemispheres, and whose ships traverse so many seas, may, whenever she pleases, easily form the most splendid and the richest assemblage of the productions of nature. The National Museum of Natural History at Paris, which is so justly esteemed superior to every collection of the kind, would not then be the only cabinet worthy of wonder and admiration; and thus a rivalry, much more honourable than that which arises from national antipathies or prejudices, would contribute once at least to the enlargement of human knowledge, and thereby to the happiness of the whole human race.*

* [It can hardly be said that the strictures here made have long since ceased to be applicable to the British Museum, which is now one of the finest in the world. It is curious that the author omits any special notice of the library of this institution which has long been one of its prominent and most valuable features.]

CHAPTER III

Arts and Manufactures.—Philosophical and Mathematical Instruments

IN England, the makers of instruments used in the sciences enjoy a deserved consideration. They are, in general, men of great information; and spare neither time nor expence to carry their workmanship to a high degree of perfection. A more careful education than is elsewhere obtainable; the demands of the navy, and the great number of persons whose wealth enables them to appreciate and to pay well for the best-constructed instruments, are causes which have concurred to form artists of high reputation, and who have served as instructors to others. I gladly embraced the opportunity of visiting several of them, under the auspices of Messrs Whitehurst and Cavallo.

I found the skilful and modest Ramsden [1735-1800] occupied in making an instrument simple in appearance, but which demanded much care and many combinations to make it perfect.

The object was to measure on the ground a base of four thousand two hundred and eighty-six toises, and to avoid the defects of the ordinary instruments of measuring; which, whether of wood or metal, are liable to be expanded by heat and contracted by cold, and to several other inconveniencies, that do not permit one to depend on their perfect accuracy, whatever precautions may be taken in using them.

To effect this purpose, it was proposed to use rods made entirely of glass; and it was in preparing these that Ramsden was then employed. The glass tubes were executed with all possible care, in the glass manufactory of Parker, to the best of my recollection. They were all of the same diameter, and straight as the most perfect ruler.

They were very long, numerous, and fixed on proper supports, with a water level to each; and could be elevated or depressed horizontally at pleasure. As these rods were to be used by placing them end to end, to ascertain the point of contact with greater precision, the extremities of each were carefully cut and ground down with emery; and on being disposed in the requisite order, they were gradually brought to touch each other by means of a screw. By this method great base-lines were obtained with a precision of measurement, of which there had previously been no example.*

I had much pleasure in conversing with Ramsden. I went to see him several times; and I purchased several instruments at his shop. He possesses all the modesty and simplicity of manners of a man of great talents.

There are also in London other able mechanicians employed in making the larger instruments of astronomy, mathematics, and physics;—such as Messrs Bird, Dolland, Adams, Nairne, Blund, Hunter, &c.

Many clock and watchmakers, who excel in their art, are likewise to be found here.

MANUFACTURES

Wedgwood

The black pottery, known under the name of *basalt*, which has the colour, the hardness, and the opacity, of the volcanic stone

* See, upon this subject, Lalande's *Astronomie*, tome iii. page 15, of the third edition, 1792; see also *Philosophical Transactions*, 1785, page 385, by General Roy and Mr D'Alby.

of that name, and its application by Mr Wedgwood [1730-1795] to busts, basso-relievos, and vases of the finest antique forms, do honour to the taste and ability of that celebrated manufacturer.

With other materials he has perfectly imitated the Etruscan vases, of which England possesses so rich a collection, thanks to the exertions of Sir William Hamilton, who procured them during his embassy at Naples, and to the public spirit of parliament which purchased them, with the view of facilitating for the English artists the study of the most excellent models.

Wedgwood has infinitely varied the art of preparing and combining the several earths, so as to form them into the most beautiful productions. He well knew that porcelains have been brought to the highest perfection in France, and that nothing can surpass those of Sêve, and some other manufactories, which have branched out from it; he therefore chose a different course, and, disdaining to be a mere copyist, he has, in a manner, created a kind of pottery peculiar to himself, and which might be regarded as absolutely original, if the vessels which come from China, of red and brown earths, without semi-transparency, and of great hardness, had not served him as models. He has, nevertheless, the merit of having excelled the Chinese, by discovering new compositions, and more especially by adopting the most elegant shapes.

As his pottery is used in every part of Europe, and as example is more efficacious than any theory or the best written instructions, it is evident that Wedgwood has contributed to a sort of revolution in the art, by multiplying agreeable forms, and accustoming the eye to enjoy their graceful proportions.

This able artist having daily occasion to study the action and different modifications of fire, has made himself, so to speak, master of that element, taking it captive, and directing it at pleasure. The course of his inquiries led him to invent a graduated instrument for ascertaining various degrees of heat, which bears his name, and does honour to his genius. The *pyrometer of Wedgwood* has a place in all the laboratories of chemistry and physics.*

* The celebrated Spallanzani has very happily applied it to determine the degree of fire necessary to fuse the lavas in volcanos; he found, that a less degree of heat than that of

But what has greatly increased the fortune of Wedgwood, and brought an immense branch of commerce to England, is his common pottery, known in France by the name of Fayance Anglaise (English-ware) and at London by that of Queen's-ware.

Its excellent workmanship, its solidity, the advantage which it possesses of sustaining the action of fire, its fine glaze, which resists the action of acids, the grace and convenience of its form, and its moderate price, have given rise to a commerce so active and so universal, that in travelling from Paris to St Petersburg, from Amsterdam to the far end of Sweden, and from Dunkirk to the southern extremity of France, one is served at every inn upon English-ware. Spain, Portugal, and Italy are supplied with it; and vessels are loaded with it for the East-Indies, the West-Indies, and the continent of America.

This universal taste, this invariable desire to procure that ware, is a sufficient demonstration that, from its solidity, form, and price, it is perfectly suitable for those who

ordinary glass-houses was sufficient for the purpose. See, upon this subject, vol. I. of his *Voyages dans les Deux Siciles*, page 14 of the Introduction and page 63 *et seq.* of the text.

make use of it. In this point of view, Wedgwood has made an excellent discovery, and has deserved well of his country, since he has given existence to an extensive branch of industry and commerce.*

Parker's Manufactory of fine Glass

Those who love the arts ought not to neglect to visit the warehouses and glass manufactory of Parker.

* France possesses all the materials requisite to imitate to perfection the English stone-ware, such as white clays, siliceous earths, minium, &c. A manufactory, established some years ago at Montereau, produces potteries, which are not, it is true, equal to those of England, but which, with some assistance and management, particularly that of employing pit-coal, would soon reach a higher degree of perfection.

Some very respectable citizens of Geneva, who commenced their labours with more intelligence and greater means, accomplished their object after repeated experiments. They made several journeys to Paris, to solicit permission to transport their establishment into France, where in the department of Isere, it would be nearer the kinds of earth which they employ; but they were scarcely listened to. I know not whether justice has been since done to their requests, or whether, weary of soliciting, they have at last renounced a project so useful for France; but this I well know, that upon every occasion where the arts are unrecognised by those who administer the finances of a great nation, the country is deprived of its resources, and compelled to have recourse to the industry of its neighbours.

In the same way it has been thought well to prohibit the importation of English manufactures; but every thing manu-

There they will see to what varied extent that substance, pure as the clearest spring water, and more delightful to the eye than crystal, may be fashioned, in the hands of skilful artists, into goblets, vases, basins, and bottles of every form.

They will be astonished at the dexterity and quickness with which the workmen cut, engrave, and polish into the purest brilliancy, articles of luxury and ornament, the

factured by them, of a superior quality, and at a cheaper rate than among us, will be imported notwithstanding, and will only become dearer than formerly. The English, who would, no doubt, wish to retaliate, will not be so ill-advised, as to prohibit the entry of our wine, which they cannot well do without. They have always subjected it to heavy duties of importation; the rich consumers pay them, and the state becomes a gainer.

I heard a man, who is well versed in this matter, say, that it were a thousand times better to permit the traffic of English merchandise, to impose taxes on it, and then to apply one-fourth of the produce, by a wise distribution, to the encouragement of French manufactures; we should, by this means, soon have as good commodities as the English. It is by such encouragement that we have been able to imitate their fine glass; it is thus that, owing to the intelligence and attention of the minister Bénézech, we have at Versailles a manufactory of fire-arms, of a finer finish and more exquisite workmanship, and also at a much cheaper price, than any made in London. Olivier, who has erected a very handsome manufactory in the Rue de la Raquette, faubourg Saint-Antoine, has imitated very well the best productions of Wedgwood. It is unfortunate, that the state of the finances has not admitted of giving this estimable artist that encouragement which he has well merited.

various utensils which cover and decorate our tables, and the beautiful lustres which illuminate them, and reflect in a thousand directions the colours and the brilliance of the prism.

Parker, like all those who are at the head of great manufactories in England, has made acquirements in more than one branch of knowledge; he has constructed a burning-lens of a large size, and of remarkable power: it is considered as the best of the kind ever made.

I had fixed a day with his friend, Mr Whitehurst, to be present at some experiments to be made with it: but the sun, as often happens to him in London, was not visible: it was not in my power, therefore, to judge personally of the effects of this large burning-glass.*

* English glass has been imitated in France: the first attempts were made, with success, in the park of Saint Cloud, at the instigation and under the auspices of Marie Antoinette, who loved and encouraged the arts. That manufactory was transported to Creuzot, near Mont Cenis, in Burgundy, where it received the name of the queen's manufactory, on account of her having contributed to establish it. Is is formed on the most extensive scale; very fine works in glass are made there; and it has already given rise to similar manufactories at Paris, and other places.

A large Brewery

It is by facts, more than by any other means, that an exact idea may be formed of what industry is capable of effecting among a people, active and animated with the genius of commerce.

A large brewery, which I visited, on the south side of Blackfriar's-bridge, excited alike my astonishment and admiration.

The buildings and yards, which are of a vast extent, have no other object than utility; every thing is solid, every thing is adapted to its purpose, but every thing is an absolute stranger to ostentation.

Seventy large horses are employed in the service of this brewery. Of a hundred workmen, unceasingly active, some prepare the malt and the hops, or are employed about the fires, the coppers, or the coolers; some rack off the beer, and others convey it into large vats, which I shall presently describe.

The beer is fermented in huge square vessels, raised to the height of the first floor; and pumps, disposed with much art, facilitate the supply of water.

When the beer is made, it descends through conduits, and is distributed, by means of pipes, into a number of casks, placed in an immense cellar. The beer becomes of a more perfect quality in those casks, where it remains, however, but a short time; from them it is drawn off by long spouts, and decanted into a great reservoir, whence it is again raised, by pumps, into vats of an astonishing size, which are placed vertically, and the top of which cannot be reached without a ladder: a gallery goes round the places which contain these vats.

Four store-rooms, on a level with the ground-floor, and of different sizes, are appropriated to receive them.

In the first, which is the smallest, there are six vats, containing each three hundred hogsheads; a hogshead contains about two hundred and forty bottles: in the second, there are twenty-eight vats, of four hundred hogsheads; in the third, fourteen of nine hundred hogsheads; and in the fourth, four of five hundred hogsheads each.

Thus their united contents amount to thirty-one thousand six hundred hogsheads.

The ordinary quantity sold, one year with

another, is about a hundred and forty thousand hogsheads. During the last war it was much more considerable, the proprietor of this brewery having had a contract for supplying the navy. One may form an estimate of the sale at that period, from the duties yielded by the beer then made. I was assured, that they amounted to ten thousand pounds sterling a month.

It was not very long since this brewery had been sold, on the death of the former proprietor; it was put to auction, and knocked down at the price of three millions two hundred and eighty-eight thousand French livres.

It is remarkable that twenty-two bidders contended for it, though it was necessary, not only to pay down that sum, but to be able to advance as much besides as would be requisite to set so vast an establishment in motion.

It is, perhaps, superfluous to observe, that almost all the beer brewed in this fine manufactory, is of the kind called *porter*, which is strong, capable of sustaining long sea-voyages, and of being preserved in bottles for many years: it is, indeed, necessary, in order to have it of a good quality, that it should remain several months in the large vats.

These vats, made of wood of the choicest quality, are constructed with an admirable solidity, accuracy, and precision, and even with a kind of elegance: some of them have as many as eighteen hoops of iron; and several were pointed out to me, which had cost ten thousand French livres a-piece.

I have already said, that they were all placed on end around the walls; but, on asking what they stood upon, my conductor shewed me, that they rested on brick arches of great solidity, strengthened by a number of thick upright pillars of wood. Their bottom was thus protected from the humidity of the ground, and it was more easily seen whether the beer escaped.

The top of each vat is most carefully covered with thick planks, joined together in the most perfect manner, and these again were covered with six inches of fine sand.

At a small distance from this brewery, there is another for making malt-vinegar, fitted up in the same manner; but, in the latter, the tuns stand in the open air, and fill an immense yard. Their size and capacity are such, that on entering the vast inclosure filled with these gigantic vats, ranged in different lines, one might imagine them, by an illusion arising from the absence of any means of exact comparison, to be a series of ships of the line, lying by the side of each other in a harbour.

The vinegar made from excellent strong beer is better than one would expect; no other kind is used throughout the whole of England, the importation of vinegar, made from wine, being severely prohibited.

Manufactory of Morocco Leather, Parchment, and Shamoy Leather

I like to see manufactures of every kind; they provide for our wants, our conveniences, and our luxury. These productions of industry are owing to the united efforts of men: they have contributed more than might be believed to the development of the human faculties; and before they arrived at that pitch of perfection to which they have been carried among some nations, the arts have gone through a thousand attempts, a thousand groping trials. Their progress is thus shown to be the same as that of the human mind, which creeps along by short steps, and travels but slowly in the path of discovery.

I like also to see the manner in which different nations pursue the same art; there are always particular processes used in one country which are not practised in another.

We are acquainted with the cause of the excellent quality of the leather made by the English; it is principally owing to the pecuniary advances which they are able to make, so as to let the hides remain longer in the pits, and to some improvements in the workmanship.

At Annonai and Grenoble, skins for white gloves are manufactured, which surpass those of England. But in England, they have for a long time had the art of making parchment, vellum, and particularly Morocco-leathers, superior to ours.

It may, indeed, be said, that the manufacture of Morocco-leather is still in its infancy in France, though it formerly had a reputation there, if we may believe a celebrated author.* But the civil wars, particularly those of religion, put to flight our finest

* Rabelais mentions the fine Morocco-leathers of Montélimar in Dauphiné.

manufactures, which were allured into Germany and England by toleration, liberty of worship, and sound policy.

I had a conversation upon this subject with a very intelligent Englishman, who proposed to conduct me to a manufactory of this article, situated at one of the ends of London, and directed by Lorrainers. I was told that I should there see a press of very great force, the effect of which was to give the desired quality to the skins to be used in the preparation of Morocco.

I visited this excellent establishment, where every thing is conducted with an intelligence and method which much interested me. But in the operations which they were good enough to let me see, I noticed nothing which I had not seen practised elsewhere. I believe, indeed, that they concealed some parts of the process from me, for they did not always reply to my questions; nor could I much disapprove of their silence. The great press, however, which is not shewn to every body, was set in motion before me, and I was made acquainted with all its details.

It is made of iron, weighs twenty-two thousand pounds, and differs but little from ordinary presses save in being larger, and of a more perfect finish. It is worked by four men, and produces a very powerful pressure; but when the highest degree of force is required, two horses are yoked to it.

Skins of different kinds, which in ordinary manufactories would have been thought sufficiently dressed, that is, which were well pressed, and shewed no trace of the fatty substances used in preparing them, were wetted and put into the press.

The water which oozed from it was collected, and during the last efforts of the press a thick oil floated on the top. This oily matter, said the manager to me, in time becomes rancid, and turns acid; it not only alters the colours applied to the skins, and gives them a blackish appearance, but also corrodes their grain, and the network which gives them consistence; so that they are thence of little durability. This observation appeared to me just, and to deserve consideration by those who are engaged in this business.

As this manufactory is considerable, a good deal of oily matter, which would otherwise be lost, is obtained by this mode, and made into tallow.

CHAPTER IV

Monument to the Fire of London.—Quakers. —Some Cabinets of Natural History.— Sir Henry Englefield.—Preparations for the Journey to Scotland, and the Isle of Staffa

THE railings of the column called the Monument were receiving repairs, which gave me a better opportunity to ascend it, and to view from the summit the city of London and its surroundings.

This column, which is situated at a short distance from London-bridge, is of the Doric order, and two hundred and two feet in height, and fifteen feet in diameter. It was erected by the city in memory of the famous fire of 1666; and is considered to be one of the master-pieces of Sir Christopher Wren.

The monument having been already described by a number of travellers, I should not have mentioned it here, were it not for a particular circumstance which my visit to it gave me an opportunity of remarking.

I had begun to ascend the three hundred and forty-eight steps leading to the upper balcony, when I perceived that the inner rail which winds round with the staircase was corroded and unserviceable. The higher I ascended the further advanced was its decay. Having reached the platform, I observed with astonishment that the balustrade which went round it, though made of iron of a considerable thickness, was almost entirely destroyed, particularly in the direction of certain currents of air; so that it would have been dangerous to approach too near it.

I presumed, indeed, that the vicinity of the sea must needs attract acid vapours, injurious to metals, but especially to iron. I observed also, that the numerous balustrades which inclose most of the houses of London, require frequent painting to preserve them. But I should never have imagined that the decay could have gone so far, in so short a space of time, supposing even that the railing of the monument had never been repaired since its foundation, that is, since the year 1666.

In several towns of the north and south of France, much nigher the sea than London, I have seen vanes of steeples, balustrades of balconies, and iron ringbolts for fastening vessels to, which although more than two hundred years old, had not sustained one-fourth part of the damage which I have noticed in the iron-work of the Monument of London.

It is hence to be inferred, that the atmosphere of this city is impregnated with corrosive emanations, more copious and active than elsewhere; and this might, indeed, be expected, where there is so great an assemblage of inhabitants, who use nothing but pit-coal, for their daily fires, throughout the whole year, and in a city filled with manufactories and works of every kind, which consume so many currents of air, and such enormous masses of combustibles.

I am very far, however, from thinking that the city of London is more unhealthy than other cities, because they only burn pit-coal here. For not only is the contrary proved by experience and a long train of observations, but it is also to be presumed that this immense quantity of fires contributes to its salubrity; in the first place, by the strong, equal, and constant heat produced by pit-coal, in an atmosphere naturally impregnated with water; and in the second place, because so many fire-places, so many manufactories and works of every kind using fire, occasion currents and changes of the air on every side, which carry off the noxious and pestilential vapours that always arise when the respirable mass is too long stagnant.

With regard to the emanations from the coal while it is burning, they are of two kinds: the first are bituminous, and even a little balsamic, and, therefore, rather salutary than injurious to the lungs. The second, which are given off when the combustible is in strong conflagration, are acid, and consequently antiseptic. But the good construction of the chimneys, and the impulsive action of the fire, raises the column of vapours above the habitations. Then the smallest wind (and there is always some blowing at a certain height) removes and dissipates these emanations, which act only on ironwork, particularly at the highest elevations, or on the foliage of trees growing too near.

Besides, the incalculable advantages yielded by pit-coal, that useful combustible, on which the existence of England, so to say, depends, amply compensate for the few slight inconveniences which attend its use: and though it might put our fashionables of Paris, like those of London, to the trouble of changing their linen twice a-day, I should wish, for the happiness of individuals, and the general prosperity of my country, that France were as far advanced as England in the general use of pit-coal.

Let not those who have only vague notions on this subject, say that we have neither so much nor such good coal as is found at Edinburgh, Glasgow, and Newcastle. To convince them of their error, I would only appeal to the opinion of several intelligent Englishmen who have travelled with advantage in France. I speak of philanthropic Englishmen, such as Arthur Young, Symonds, Sir W. Hamilton, Lord Greville, and others; for, with respect to the British government, its policy too urgently requires that we should long continue ignorant on a point of so much importance, which so closely affects our manufactures and our commerce. I shall return to this useful subject, which I quit with regret, when I shall have visited the mines of Newcastle.

The Quakers

I love the Quakers, and I have great pleasure in seeing them in private, in society, and in their religious assemblies. They inspire me with an involuntary veneration.

Clothed with all that is most simple, plain, and modest, but at the same time, most neat, finished, and perfect, it has seemed to me that their mind shares in the whiteness of their beautiful linen, and must be as pure and as carefully tended as their clothes.

Buffon had good reason to say, that men become, so to say, identified with their attire, and that it is of much more importance than might be imagined, to accustom youth to attend to what is called a befitting and decent dress and deportment. There is much profound sense in what that illustrious man wrote upon this subject. He has said further "We form our opinion of a man from his physiognomy; there is nothing about him, even to his clothes and to the way he dresses his hair, which has not some influence on our judgment. A sensible man should regard his clothing as forming part of himself, since it does so, in fact, in the eyes of others, and since it counts for something in the whole idea which one forms of the person who wears it." *

The places where the quakers assemble

* *Histoire Naturelle de l'Homme, page* 520, *in* 4*to.* A man who was not destitute of talents, but who wished at all cost to act a conspicuous part at too early a period, I mean Hérault de Séchelles, made an excursion to Montbart, in 1785, to see Buffon, who kept him several days at his house, and was pleased to shew him marks of affability, and even of confidence.

In gratitude for so many proofs of kindness, Hérault, on his return to Paris, made the greatest haste to announce that he had filled his journal with anecdotes and lively details, respecting the great man whom he had been to visit; and he read in more than one circle, with a tone of mystery, this journal, written under the roof of hospitality. This composition, loaded with the minutest details, is an heterogeneous mixture of pompous eulogies, critical remarks, and satirical episodes, often scandalous and almost always false. This species of domestic spying, which so strongly savours of ingratitude and bad manners, would, in other times, have driven that man out of every society that had any respect for itself.

Hérault appears to have wished to throw ridicule on Buffon on account of his hair, bleached with the labours of sixty years, of which he was very careful. He has made a parade of saying, that Buffon "caused it be dressed three times every day in five rows of floating ringlets."

The following is a truer fact upon this subject, of which I was a witness. Madame de Nanteuil, a woman of great vivacity and beauty, came one day to see Buffon at Paris, about eight in the morning, on some business. He was at his toilette; she apologized for appearing in deshabille before the historian of nature.—"What!" replied Buffon, "are you not already sufficiently adorned with youth and beauty. It is at my time of life that attention should be paid to one's hair and one's dress, in order to conceal a little the deformities of age."

for worship, or rather for meditation, where they descend into their own thoughts, and await the inspirations of virtue in their hearts, are calculated to awaken respect.

This kind of temples, like those of the nations of antiquity, admits the light only from the roof. The walls are of a dazzling white; the wainscotting, unencumbered with sculpture, shines in the modest lustre of its native colours, and the exquisite cleanliness with which it is kept; the seats are simple benches, placed in parallel rows. In vain would one look here for paintings, statues, altars, priests, and acolytes. All these accessories are considered by the quakers as excrescences, devised by man, and foreign to the Supreme Being. They prefer to offer up to him pure hearts, and acts of virtue and beneficence. They are persuaded that nothing can be more agreeable to him than that mild philanthropy which induces them to regard all men as brothers and real friends, with whom they travel, in common, the short, but difficult road of life, in which they reciprocally stand in need of assistance.

They, accordingly, hold in abhorrence those cruel and sanguinary persons, who, from motives of ambition or vengeance, provoke war; that is, compel or excite men who have no real cause of quarrel, to cut each other's throats in cold blood.

When the quakers are assembled in their churches, the men occupy a place apart from that of the women, and have their heads covered with a black hat, of a broad half-cocked brim, without loop or button. Their eyes are humbly bent on the ground, and often entirely shut, to avoid any distraction in the midst of their contemplative meditations.

The women also have their heads covered with bonnets, made of silk, velvet, or straw, but very plain. They, in general, conceal their faces; at least they do so in this place of meditation. Their hair, too, is without powder, but is washed and trimmed with such neatness, that it forms one of their finest ornaments. They are attired in the most decent taste; their clothes, however, are generally made of the finest and choicest stuffs, though at the same time in a style of the greatest modesty.

At the farther end of the church there is a kind of platform, a little raised, and surrounded with a wooden balustrade. It is not a pulpit; it is rather a large and long tribune from which to harangue. Here it is that those (including the women) who are animated by heavenly inspiration, take their place, to communicate in a loud voice to their brethren the transports of their souls, and the impressive thoughts which the Eternal has sent to them.

I have often beheld them, in that prophetic state, with a perfect conviction that they were no more inspired by the breathing of the Holy Ghost, than was the Sybil in the time of the Oracles by the spirit of Apollo, or the Somnabulists, lately, by the illusions and signs of Mesmer.

But, making a distinction between the quakers, who are certainly sincere in their belief, and the latter, who exhibited nothing but imposture, I liked to trace the impressions producd on them by the action of the mind, when too long employed in metaphysical abstractions. I saw some who, from fatigue of the head, ended by forgetting themselves, and believing that they were inspired, broke the profound silence which reigns in these assemblies, and mounted the tribune.

Then it was that I delighted to watch

them more closely, with the help of a glass. They keep their eyes half shut, or bent towards the ground, while slowly, and at long intervals, they pronounce some words in a sad and melancholy tone; supporting themselves with their hands strongly pressed against the balustrade of the rostrum, and seeming to make efforts to reach, and, as it were, to seize thoughts.

They then sway themselves backwards and forwards, sometimes sideways; at first with a slow and uniform motion, uttering some words more rapidly. Their action then redoubles: and this struggle of body and mind soon drives the blood towards the head; the cheeks redden, a crowd of thoughts arises, expressions follow; the whole soul and heart are kindled; a sort of quaking appears, and the orator is inspired.*

The women in similar circumstances follow nearly the same course as the men; they are neither more nor less loquacious. Many of these discourses are below mediocrity; some are tolerable; it is even said that some are very eloquent; but all of them are favourably received. The subject always turns on the

* It is this which has procured them the name of *quakers*.

duties of man, on the pardon of sin, and on lessons of the most perfect morality. I heard a woman, one day, improvise a very fine prayer to God. She might possibly have known it by heart beforehand; or her feeling soul had inspired that fine emotion of love and gratitude. Women will always give us lessons in this respect.

There are meetings at which nobody speaks, as might be expected among men of worth, happy in their own consciences, and more accustomed to put morality in practice, than to cry it up in words. As there are here no bespoken discourses, nor a pastor who governs the flock at pleasure, the only rule observed in this matter is, never to speak but from the impulse and transport of the heart. But as these depend upon an ardent soul, and a strong imagination, under the influence of physical causes, which must vary according to the season, the state of the air, or that of the health, it follows that the thermometer of the head, no less sensitive than that of art, must sometimes be found in a state of immobility and stagnation.

But what seems to distinguish this simple worship from many others is, that experience has proved that it leads men to the practice of their duties without wearying them with vain superstitions; that it makes virtue amiable, by presenting her under attractive forms; and that men of this disposition are valuable to a government, by their good example. Happy in their good deeds; rich, in general, from their application to industry, they furnish the strongest proof that the morality of individuals, by creating private happiness, gives birth to public prosperity.

Some Cabinets of Natural History

Mr Drury's cabinet of insects has been formed at a great expense. Much time and many fortunate opportunities must have been required, in order to collect so many rare objects of this kind, brought from the East-Indies, from China, Japan, the South Sea, &c. Every thing is arranged, in this collection with much care and great neatness.

Smeathman, who had travelled in Africa, and who brought home several curious insects to Mr Drury, with whom he was particularly acquainted, procured me admission to this cabinet. This recommendation was all the more useful to me, in that Mr Drury had the complaisance to show me his collection in minutest detail, and with much affability.

I knew, and was possessed of, his excellent work, entitled, *Natural History of Insects, classed according to their different Kinds, in English and French* (*3 vols. in 4to, with coloured plates*; 1770, *and following years*). I saw, therefore, with much interest, the insects which were delineated in this book.

I passed also some hours very agreeably in the cabinet of Mr Thomas Sheldon, brother of the anatomist. It contains South-Sea shells, and other interesting marine productions.

I infinitely regretted not to have seen the rich mineral collection of Lord Greville, the relative of Sir William Hamilton, ambassador at Naples; but they had both a little before set off for Scotland, with the intention of visiting the Isle of Staffa.

I was also deprived of a sight of the scientific collection of Lord Bute, as well as that of Dr Pearson; neither of them being in London at the time.

Sir Henry Englefield compensated me for these privations by the civilities and kindnesses which he heaped upon me, during my stay in that city. He has successfully applied himself to the study of astronomy and physics; he is an agreeable companion, full of affability and worth, and gives a hearty welcome to strangers. I sincerely wish to have the pleasure of seeing him some day in France, that I may repay with heartfelt gratitude the civilities I received from himself as well as from his estimable mother.

If all Englishmen were endowed with such urbanity, it would be unjust to reproach them with that neglect and coldness which they are accused of showing towards those who have given them the best reception in France. This accusation, however, is exaggerated; and I have more than one proof that there are many exceptions to it.

As I wished to take advantage of the remainder of the good season, for my journey to Scotland and the Hebrides, I was busied for several days in making the necessary preparations for my departure.

Several learned men were pleased to give me letters of recommendation for Glasgow and Edinburgh, and to the Duke of Argyle, who was then in one of his estates in the north of Scotland, on the road which I had to take to the nearest point of embarkation for the Hebrides.

CHAPTER V

Departure for Scotland.—Itinerary.—Observations of Natural History

COUNT Paul Andreani, of Milan, William Thornton, M. de Mecies, and myself, set off from London at six o'clock in the evening, in three post-chaises; two of which were occupied by ourselves, and the third by our servants.

I had known Count Andreani at Paris; he loved the sciences, and had made a very fine aerostatic experiment in Milan, at his own expense; he went up in a large balloon, which he caused to be constructed on the plan of Montgolfier.

William Thornton is a very worthy and intelligent American, who, after prosecuting his studies with advantage, under Doctor Cullen, at Edinburgh, had come to finish them at Paris, where he conceived a taste for natural history. The journey could not but be very agreeable with such pleasant companions.

M. de Mecies, of London, had been introduced to us a few days before our departure from London, by Mr Thompson, a very good naturalist, as a studious young man, who was much attached to mineralogy; we admitted him, with pleasure, into our party.

It was pleasant to associate with persons who had the same tastes as ourselves, and were not afraid to share in the fatigues and dangers of the tour which we meant to pursue as far as the isle of Staffa, if the season should permit us to risk ourselves on the tempestuous sea which surrounds it, and which is scattered over with islands and dangerous currents.

Itinerary

From London to *Barnet*, twelve miles.—A superb road, covered with carriages, and with people on horseback and on foot, who were returning, in a fine moonlight evening, to London, from the country-houses and neighbouring villages, where they go to recreate themselves during Sunday.

We found the air so serene, and the night so delightful, that we resolved to profit by it.

Hatfield, nine miles.

Stevenage, twelve miles.—We arrived here at four in the morning, and rested till nine. Inn excellent, but very dear.

Dugden, sixteen miles.

Stilton, fourteen miles.

Nothing can surpass the beauty and convenience of the road during these sixty-three miles; it resembles the avenue of a magnificent garden.

At Stilton, one begins to observe, on the sides of the road, large heaps of stones, destined to repair it.

These stones are calcareous, and of a greyish colour. They contain a number of petrified marine shells, among which I distinguished a kind of *Concha exotica*, and others of a more common sort. In these stones, also, by the side of the shells, may be seen pieces of well-preserved wood, which it is difficult to keep, because being of a pyritous nature, they are easily decomposed by the air. They are of a black colour, and some bituminous parts may be distinguished in them of greater solidity.

On leaving the village of Stilton, I observed, at the door of the last house, on the right, in the way to Stamford, a sort of seat of unhewn stone, consisting of a block of true, black volcanic basalt, mixed with some small crystals of black schorl, and specks of volcanic chrysolite. I asked several persons whether they knew whence that stone had come; but was able to procure no other information than that it had always been seen in its present place, and that they did not know whence it had been brought. As it weighs, however, at least two hundred pounds, and as it is probable that it will not be removed for a long time, I invite the English naturalists, if there be any in the vicinity of Stilton, to discover the place where this volcanic stone was found, and to examine whether it has come from the neighbouring mountains.*

From *Stilton* to *Stamford*, fourteen miles. —There are two old churches at Stamford, which are worthy of being examined. Their construction is solid, and at the same time, bold and elegant. The architecture is a simple Gothic not without merit; and the

* [Blocks of this kind occur abundantly in the Boulder-clay of the eastern counties, and are scattered over the surface. They have been carried from various northern sources during the Ice Age.]

execution is otherwise such as leaves nothing to be desired.

Wintham Common, eleven miles.

Grantham, ten miles.—A superb inn, of exquisite neatness.

Newark, *South Muscomb*, *Tuxford*, fourteen miles.—Road less agreeable, through parish meadows, a little marshy. In some parts, however, there may be observed, under the turf, beds of black limestone, which break up into layers from six lines to four inches thick. This stone, when rubbed with iron, emits a smell of burnt horn. There are found in it terebratulæ and other small shells.

Barnby Moor, ten miles.

Doncaster, fourteen miles. A handsome little village. Not long before this time a balloon had been sent up here, filled with rarefied air, after the manner of Montgolfier. I saw an account of it in a bill at the door of the posting-house.

Ferrybridge, fifteen miles.—A continuation of communal pastures from Barnby Moor to Ferrybridge, where numerous flocks of sheep, and herds of cattle, may be seen, as well as many horses. The soil below the meadows consists of small gravel, in some places covering beds of calcareous stone. On approaching Ferrybridge, the country becomes hilly, and considerable banks of grey limestone are to be seen. At Ferrybridge a good inn for post-horses is kept by J. Denton.

Brotherton, *Fairburn*, *Micklefield*, *Aberford*, *Bramham*, *Wetherby*, *Walshford Bridge*, *Allerton-Park*, *Boroughbridge*, *Dishforth*, *Topcliffe*, *Busby Stoop*, *Sand Hutton*, *South Ottrington*, *Northallerton*.—The same order of things, with very little difference. Road practicable, but not so good as before; landscape rather wild; some parts more populous and better cultivated.

The face of the country is intersected at Northallerton with hills, for the greater part consisting of, or covered by, large round pebbles. At intervals, however, and in the ravines, strata of a greyish-white calcareous stone of rather indifferent quality make their appearance. But the stone is used with advantage in agriculture, by burning it into lime, with which they manure the land.

As for the rounded blocks of stone, which cover the most of these calcareous hills, they proclaim a new order of things. They include granites, greenish petrosilex, and many black traps, which it would be difficult not to confound with compact volcanic lavas, if one's eyes were not very much accustomed to stones of this kind.

Lovesome Hill, *Little Smeaton*, *Dalton*, *Croft*, *Darlington*.—On leaving the little town of Darlington, we saw by the side of the road considerable heaps of black traps, which had been brought from some places in the neighbourhood, to keep the road in repair.

Coatham, *Mundeville*, *Aycliffe*, *Woodham*, *Ferryhill*, *Sunderland-Bridge*, *Durham*.—The last a small city, the see of a bishop, placed in a delightful situation, with a superb Gothic cathedral.

Durrowmoor.—Here traces of coal-mines begin to appear in a poor and partly argillaceous calcareous stone.

Plawsworth, *Chester-le-street*, *Pelaw*, *Birtley*, *Gateshead*.—Coal-mines worked at Gateshead.

Newcastle.—From Ferrybridge to Newcastle is reckoned ninety-six miles. We made this long journey in one day; having left Ferrybridge, where we had passed the night, at exactly five in the morning, and reaching Newcastle at nine in the evening.

About four in the morning, of the 30th of August, when advanced about seventy miles on our way from London, we began to feel the weather cold and sharp, though it was at the same time calm and serene, and the air quite pure. I inspected my thermometer, and found it half a degree below the freezing point; I saw ice, also, of about half a line thick. At the same hour of the following day, the mercury stood ten degrees higher, and continued so almost the whole day.

On the second day, it was fifteen degrees above the zero of Reaumur's thermometer, and the cold had then disappeared. This shews a great inequality of temperature at this season in England, where the winter is a little longer and more foggy than at Paris; but where it is less cold, on account of the vicinity of the sea.

CHAPTER VI

Newcastle.—Its Manufactures.—Its Coal-mines

NEWCASTLE is situated on the beautiful river Tyne, which is covered with vessels, and bordered on the right and left with manufactories of every kind, down as far as its mouth, which is about ten miles from the city.

I stayed there long enough to enable me to study its numerous coal-mines, and the manifold products of the most active industry.

Mr David Crawford, who was the friend of William Thornton, one of my fellow-travellers, procured us access to the mines, and to several manufactories: he did so with all the more zeal and kindness, in that he himself loved natural history and the arts, so that he was very communicative, and took the greatest trouble to show us every thing that was curious. He was the proprietor of a manufactory, entirely devoted to the extraction of gold and silver from the cinders of the workshops of mints, and those of goldsmiths and gold-wire-drawers, as well as from the debris of crucibles and cupels. He purchases these raw materials in Holland, England, and France.

But it is remarkable, that the cinders brought for this purpose in large quantity from France, have already passed through the hands of refiners, who use only washing and other imperfect processes, by which means they recover only a part of the precious metals. In Newcastle, however, from the abundance and cheapness of coal, the materials can be treated, by fusion, in reverberating furnaces, very ingeniously constructed for the purpose.

I saw, with much interest, the manufactory of Mr Crawford, where he has erected other furnaces for the revivification of the calx of lead and copper. He procures the materials for this operation from different parts of Europe, by purchasing old leaden pipes, which have remained long under the ground, copper which has been corroded by rust, and worn-out cannon which he obtains at a low price.

We saw a number of manufactories, where

window-glass, bottles, decanters, and drinking glasses, are made. All these works, though established in buildings with hardly any appearance, are equipped and directed with a simplicity and an economy worthy of notice.

This modest simplicity has the great merit of encouraging active and industrious men, who would otherwise be unwilling to form establishments, being discouraged by the initial expenses which extensive constructions entail.

It is almost always ostentation and vast buildings which in France ruin manufactures, and prevent those of which we stand in need from being established: men are too afraid of involving themselves in ruinous expense for warehouses and workshops.

It must be acknowledged, that the English and Dutch are more prudent, and give us in this matter lessons, which we should do well to imitate. Architecture is a plague in establishments of this kind.

This beautiful river Tyne displays along both its banks a crowd of manufactures which give it a very striking appearance. On one hand are seen brick-fields, potteries, glass-houses, earthenware-works, and works for making white-lead, minium, and vitriol; on the other, manufactories of sheet-iron, of tin, of all kinds of utensils, brass-wire mills, flattening mills, &c.

From this multiplicity and variety of establishments, rising opposite to one another, there is diffused every where so much activity, so much movement, and if I may use the expression, so much life, that the eye is agreeably surprised, and the mind feels a lively satisfaction in contemplating such a magnificent picture, wherein so many useful men can be seen to find ease and happiness in work, while at the same time, they promote the well-being of others; and, as the last result, they contribute to the prosperity of the government, which watches over the safety of all.

Compare this honourable industry with the ignoble indolence and disgusting misery of that crowd which, calling itself Roman Catholic, throngs the doors of the churches and monasteries in countries governed by pernicious laws, and it will be seen whether or not governments influence the happiness of mankind.

The coal-mines in the neighbourhood of Newcastle are so numerous that they may be regarded as not only one of the immense magazines of England, but also as the source of a profitable foreign commerce.

Vessels loaded with coal, for London and different parts of Europe, sail daily from this port, and, so to say, every hour of the day. Besides this commerce, the navigation which results from working these mines, gives an incalculable advantage to the navy, by forming a great nursery of seamen. In time of war, more than a thousand coal vessels can be armed, and do considerable injury to the enemy's commerce.

In this practical school of navigation are to be found men inured to every danger. The celebrated Cook began his naval career, as a sailor, on board of a Newcastle collier; and his capable and active genius soon raised him to be skipper. He afterwards purchased a ship on his own account; on occasions of danger he knew so well how, as it were, to master the elements, that though yet young, he acquired a great reputation among his brother seamen. His high qualities eventually obtained for him so completely the confidence of the English government, that this wonderful navigator sailed three times round the world, enriching geography, natural history, and navigation, with the greatest discoveries. The modest house in which he was born, in the neighbourhood of Newcastle, is preserved with veneration.

The coal-mines, in the neighbourhood of Newcastle, are situated in so fortunate a position that the soil which covers them yields fine pasture that supports herds of horses. Under this fertile soil there is found a sand-stone, of excellent quality for grind-stones. This second richness of the earth forms another extensive object of trade for the industry of the inhabitants of Newcastle: these stones have so great a reputation, that they are exported to all the ports in Europe.

The first mine I visited belongs to a private individual; it is situated about two miles from the town, and requires one hundred men to work it; thirty for the work above ground, and seventy in the pit: twenty horses live in this profound abyss, and drag the coal through the subterranean passages to the pit-bottom; four outside work the machine which raises the coal, and some more are employed in auxiliary labours.

The following is the order of the mineral substances, as they appear in descending to the coal:

	Feet.
Vegetable soil, of good quality . . .	2
Beds of rounded pieces of limestone and sandstone	15
Grey clay, more or less pure . . .	16
Hard quartzose sandstone, with flakes of mica	25
Very hard black clay, somewhat bituminous, intermixed with some specks of mica . .	26
Black clay, more bituminous, and partly inflammable; when the laminae of this clay, which separate with facility, are examined with attention, some prints of ferns appear, but they are scarcely discernible . . .	18
Total	102

At this depth of one hundred and two feet the coal is found. The seam is five feet thick in some places, and varies in others; but in general it is easily wrought, and much of it is brought up in large blocks. This last circumstance is of considerable advantage, as such pieces are always easily transported, and are besides well suited for chamber-fires; which makes this kind of coal sell at a higher price.

When the bed of black and bituminous clay is penetrated, the coal is found adhering to it; but this is not always the case, for there are other mines in the neighbourhood where the roof is of sandstone, which in the points of contact is mixed with the coal to the thickness of two or three inches; the latter imbedded in the sandstone, in the form of splinters which, when attentively examined, have the appearance of wood.

This mine has a large steam-engine for pumping out the water, and at the same time working a ventilator to purify the air.

The winding machine which raises the coal from the pit is convenient, and easily worked by two stout horses. The buckets, in which the coal is brought up, are not of wood, but of osier, strongly made, and having an iron handle. They contain at least twelve hundred pounds of coal each; and as the one ascends while the other descends, one of these baskets arrives at the mouth of the pit every four minutes. It is received by a single man who while it is yet suspended, places it upon a truck drawn by one horse. He then unhooks the basket, puts an empty one in its place, and pushes the truck to a place somewhat raised at a short distance, where he empties the basket on the latticed roof above a kind of shed ; the dust passes through the open spaces and falls below, while the large pieces of coal rolling down the inclined plane, fall upon the ground in heaps on the outside of the shed. Waggons, which I am about to describe, then take it up, and carry it to wharfs on the river-side.

It might be expected that the land transport of such an immense amount of coal would require numberless horses and men, which would involve immense expense. But art has surmounted this difficulty in the following manner.

Roads which have an almost insensible inclination are formed with the greatest care, and prolonged to the place where the vessels are loaded. The length of these roads is often more than several miles.

This first operation being finished, two parallel lines are traced along the road, at the exact distance which separates the wheels of the waggons. Logs of hard wood are then laid along these two parallel lines, and firmly fixed in the earth with pins.

The upper surface of these logs is carefully cut into a kind of moulding, which is well rounded, and projects upwards. The thickness of this elevated ledge must correspond with the width of the groove in the waggon-wheels, which are made of cast-iron, and hollowed in the manner of a metal pulley.

These wheels are completely cast in one piece, in a mould from which the rim comes out hollowed. This large groove is several inches deep, and of a proportional width; so that the wheel exactly encases the projecting part of the log, from which it cannot slide in any direction. As the moulding is well greased and is also polished by continual friction, four-wheeled waggons, containing eight thousand weight of coal each, move along the inclined plane, by the laws of gravity, and proceed as it were by magic one after another, until they reach the Tyne. Arrived there, a strongly and artistically made wooden frame prolongs the road for several fathoms at such a height above the water as to permit vessels to pass below it on lowering their masts. A man stationed on the platform opens a hatch, whence a large wooden hopper descends towards the vessel, the hatches of which are open. When the waggon comes to the trap in the platform it stops, its conical bottom opens, and all the coal runs in a moment through the hopper into the vessel. The waggon being emptied, returns by a second road parallel to the first. Other waggons follow the same course after having been in this manner relieved of their contents; and in a short time the vessel is loaded. A few horses suffice to bring back the empty waggons to the pit, and they soon return with a new freight of coal. This contrivance, as expeditious as it is economical, soon repays the cost of constructing such roads.

I have here given but a rapid sketch of these extraordinary roads, which are varied in several ways. It would require me to enter into details which might prove too long, and ill-suited to the nature of this work, were I to describe all the ingenious means which art and industry have employed in working wonders of this kind. Where local circumstances have permitted, the weight of the load, and the accelerated movement have been combined in such a manner, that files of loaded waggons run down the inclined plane and at the same time cause the empty waggons to reascend without the assistance of horses, along another road parallel to the first.

The great economy produced by these ingenious contrivances, which save the employment of men and horses, enables the English to sell the coal which they export in such abundance to all our ports on the ocean and the Mediterranean, at a price lower than that of our own mines, in all cases where we have to bring it only three or four leagues by land. Marseilles affords an example in point. This town, which consumes immense quantities of fuel in its great soap manufactories, is within four or five leagues of abundant coal-mines. This coal is indeed of an inferior quality, but it is nevertheless used with advantage in the furnaces of soap-works. Would any one believe that the coal of England, which is excellent, lasts double the time, and gives double the heat, when sold duty free in the port of Marseilles, is cheaper than the former. Such instances as this ought doubtless to give us very important lessons.

The industry of the inhabitants of Newcastle is so active, that accustomed to apply it to every thing, they have even turned to profit the pyrites, which injures the quality of the coal, but which is found in great abundance in some of the mines. The pyritous substances are carefully separated from the coal; and the expense which this labour may occasion is repaid with usury by the vitriol which is produced. The process by which the vitriol is extracted is at once simple and economical, and does honour to the intelligence of those who first put it in practice.

A large area of ground is enclosed, to which a gentle but sensible declivity is given. The surface made even, and well beaten, is then covered completely with an unctuous clay, which is smoothed as if it were plaster, in order to prevent the water from percolating into the ground. At the same time a furrow is formed in the midst of the area, calculated to collect all the water towards one point, and to convey it to a reservoir.

The area being thus prepared, the pyrites is spread all over its surface in layers one above another, to a thickness of several feet, care being taken, in placing the different pieces, to leave intervals for the access of air.

This mass of pyrites, exposed to the changes of the atmosphere and the seasons, soon heats, swells, and effloresces. The operation is accelerated by occasionally turning the pyrites with iron rakes having long teeth, so as to expose new surfaces to the air.

In a warm and dry summer, it is frequently necessary to sprinkle the mass of pyrites with water. This answers two purposes; first, to wash away the salt which has already formed; secondly, to set up by the moist heat a kind of fermentation in the pyrites, which gives rise to a quicker decomposition. Gentle showers are therefore also excellent for forwarding this operation. Finally, the water which is loaded with vitriol, finding a clay bottom which prevents it from losing itself in the earth, flows down the inclined area, and falls into a reservoir, where it clarifies. The natural evaporation which takes place adds to its strength; and when it is judged to be ready, it is drawn off into a second reservoir, attached to the work-houses of the manufactory: from thence it passes into leaden cauldrons, into which old rusty iron is thrown. It is then made to boil, and evaporate by a strong fire, formed of an inferior kind of coal; and when the liquid is sufficiently condensed, it is cooled and crystallized in wooden troughs. Two or three men are sufficient to manage a large manufactory of this kind, and to make a considerable quantity of copperas.

France formerly bought a great deal of this Mars vitriol [iron sulphate]. The dyers of Rouen, Paris, Lyons, and Marseilles, consumed astonishing quantities of it; but they now procure it nearer home, since there have been established near Alais, in Languedoc, two manufactories, which make this salt almost in the same manner as those of England, except that the French pyrites is not obtained from the coal-formation, but from a thick vein which traverses a grey limestone. Establishments of this kind might be increased to a great extent in France, where pyrites abounds in many places; but care ought to be taken to erect them in the neighbourhood of wood or coal, and above all, near rivers, that the advantage of water-carriage may be obtained.

Moreover it is mere prejudice to think that the English copperas is better than that of France. There is no difference between them when they are made with equal care. It is to be wished that we might relinquish a number of errors of this kind, which are founded solely in habit and custom, and are always injurious to the interest and commerce of a country.

The great quantity of coal-dust that comes from more than a hundred coal-pits at work in the neighbourhood of Newcastle, would soon become a serious incumbrance, were it not that an admirable method has been fallen upon to prevent the inconvenience, by preparations as simple as they are ingenious. Coal in this pulverized state is not fit for household use, because it falls through the bars of the grate, and puts out the fire. It is only good for glass-houses, lime-kilns, forges of blacksmiths and farriers. The consumption for these purposes is indeed very considerable, but is not nearly equal to the quantity produced by the pits, notwithstanding the great care that is taken to extract the coal in large pieces: besides, some kinds of coal, on the least blow, are liable to crumble into fragments. Means have therefore been sought to make this coal available for use in grates.

At Liège this kind of coal is kneaded with clay, and formed into balls, or a sort of bricks, which when dried are hard enough to be used in stoves, and even in grates. This contrivance, however, cannot be put in practice save in countries where labour is cheap; such as Brabant, where the preparation of the coal is assigned only to women, accustomed from youth to perform this troublesome and disagreeable task.

But this method would not answer in a rich country, abounding with coal-mines, and where the means employed for facilitating the working and the carriage of the coal are upon the most extensive scale; besides, the comfort and cleanliness of a private house in England would hardly be compatible with the use of a fuel which every moment covers the hearth with earthy cinders and dust.

It was therefore necessary to seek for a method more in harmony with the habits of the people, and more reconcileable with the dearness of labour in England.

The tendency of the best kind of coal to cake together into a single mass, as it goes on burning, naturally suggested the idea of endeavouring to consolidate, by means of a strong heat, considerable quantities of this coal-dust.

It appears that, as far back as the year 1682, Becher, a German chemist, gave the first hint to the English on this subject. He not only proposed to remove the disagreeable smell of mineral coal, by converting it into a kind of charcoal resembling that of wood, but to extract from it by the same operation a sort of tar, which he regarded as superior to that of Sweden. He tells us himself that he made a most successful experiment to this effect in England.* But I have elsewhere shewn, that though the process used by Becher was very ingenious, it presented many difficulties in the execution, and could not be carried into practice upon a large scale: besides it would cause a great waste of coal.

Since that period, more simple and expeditious means have been discovered. The coal-dust, put into a kind of oven, which

* See the German work by Becher, entitled *Närrische Weisheit, und Weise Närrheit* (Foolish Wisdom and Wise Folly), printed at Frankfort in 1683, in 12mo. See also the work which I published in 1790 under the title of *Essai sur le Goudron du Charbon de terre, etc. précédé de Recherches sur l'origine et les différentes espèces de Charbon de terre.* Paris, Imprimerie Royale, 1790, in 8vo.

is in the first place well heated with large pieces of coal, cakes together without losing any thing except its bitumen. When the mass is incandescent and quite red, large pieces of it are pulled out with iron rakes, and laid on the ground, which they scarcely reach before they go out. They are solid and spongy, and are used with great advantage, not only for chamber-fires, but what is much more important, for smelting iron ore, in high furnaces. This ingenious contrivance has given birth to several new branches of industry and commerce.

The coal thus prepared is called in England *coke*, and is used in a great number of manufactures, as a substitute for wood-charcoal, to which it is in several operations superior, seeing that it gives a stronger, more equal, and more continuous heat.

The same process has been employed in France, where the coal-dust is converted into a substance resembling charcoal as in the English manufacture, but with some useful improvements.

The coal thus prepared is called in France

Charbon épuré, or *Charbon désouffré*. The city of Paris consumes it in great quantities, which are prepared at Moulins en Bourbounois, at Saint-Etienne en Forest, &c., and which are transported by the Loire, the Allier, the Canal de Briare, and the Seine. This admirable means of preserving our great and valuable forests is a thousand times more efficacious than that crowd of laws, of regulations, and of employés, which tend only to destroy them.

The city of Lyons has likewise a fine establishment of this kind, situated near the Pointe d'Enée, * and the *Charbon épuré* prepared at Saint-Chaumont and Rive de Giers is used in the copper-works of Saint-Bel.

But, thanks to the government, and to a rich and enterprizing Company, an iron foundry, which will soon rival the best works of that kind in England, is established at Creuzot, near Montcenis, in Burgundy, in a

* Since the Revolution not one bushel of *Charbon épuré* [refined coal] has been brought to Paris. The wood of our finest forests is there reduced to ashes. The establishments of Moulins, Lyons, and Rive de Giers, have disappeared, and the worthy men who erected the foundry of Creuzot, have almost all watered the earth with their blood.

place originally sterile and solitary, but now covered with habitations. The abundance of coal, the simple mode of preparing it as a substitute for wood, and the models furnished by the ingenious Wilkinson, have, so to say, worked miracles, and have given birth to an establishment which is truly worthy of a great nation.

From this short sketch it will be seen how incalculable are the advantages of every kind which the use of coal confers on human society.

I insist on this truth because a country having so wide an area, and so large a population as France will be compelled, when its forests have been consumed, to resort to another kind of fuel. Fortunately it possesses numerous deposits of coal, the greater part of which has not yet been opened up, and the fine rivers and streams which traverse the country afford every facility for the formation of canals. It will soon be full time that attention should be given to the efficiency of this resource. To individuals it would give comfort and happiness, and to the government a source of prosperity not yet even suspected.

I was one day at Benjamin Franklin's

house at Passy: several Americans of exceptional ability, who were intimately acquainted with the political and commercial state of England, were of the party. I shall not name them, because they have since played a distinguished part; but I heard them, with pleasure, say that no publicist has recognised the real cause that has made the destiny of England so prosperous. "It cannot be doubted," said one of them, "but that it is the coal-mines which work so many miracles: we know that it is almost a national treason to say so in France, where the coal is as plentiful and as good as in England; but the French have supported the liberty of the United States; and besides, I wish to see the people of every country happy. I have travelled much in Italy and in France, and in passing through the latter country, in the midst of winter, I observed, with great sorrow, that in several provinces most of the inhabitants of the country-districts and even those of some towns suffered grievously from want of fuel. The effect of the cold was such that whole families were compelled to remain in bed, in a state of stupid torpor, which deprived them of the fruits of their toil, and consumed, in a few days, all their little savings. How different is the case in England, where the winters are much longer, though less severe, than in the north of France! The peasants, beside a large coal-fire, which, at the same time, lights and warms the cottage, are happy and satisfied. The father prepares his implements of husbandry for the ensuing spring; his sons make nails or other objects; his daughters spin wool or cotton; the mother is busy with the cares of the family; and as the coal-fire is kept up all day and a part of the night, their work is prolonged in ease and comfort. The manufactories of every kind, whether in town or in country, always manifest the same activity. As nobody suffers from cold, people are exempt from most of the illnesses of winter: thus, that season, usually so fatal to others, scarcely at all diminishes the labour of this fortunate people. There necessarily results from this multitude of workers, ceaselessly active, a mass of wealth, equally advantageous to the state and to individuals, who owe this condition of comfort to coal."

These words, full of truth and judgment,

made so deep an impression upon my mind that they have never faded from my memory, and they served to direct my special attention to the coal-mines of happy England. Doubtless they similarly interested the illustrious man in whose house this conversation passed, and who was a thousand times better qualified to feel and appreciate them than I was. The following is an extract from a letter upon this subject, which he shortly afterwards addressed to an estimable and learned friend, who has honoured me with several marks of kindness: "Wood will become extremely scarce in France, if the use of pit-coal be not introduced in that country as it has been in England, where it at first met with some opposition. In the records of parliament, during the reign of Queen Elizabeth, it will be found, that a motion was made, that *several dyers, brewers, blacksmiths, and other artizans and manufacturers of London, had taken to the use of coal for their fires instead of wood, whereby the air was filled with unwholesome vapours and smoke, to the great prejudice of the health of the inhabitants, particularly of persons who had lately come from the country; and it was proposed* *that a law should be enacted prohibiting these artizans from using this kind of fuel, at least during the session of parliament.* From this it appears, that coal was not then used in private houses, because it was considered unhealthy. But, fortunately for the inhabitants of London, they have paid little attention to this objection; and they now believe that coal rather contributes to make the air salubrious. Indeed, since its use became general, they have not been subject to those pestilential fevers which formerly so severely afflicted them. Paris is put to enormous and constantly increasing expense in the consumption of wood, because the inhabitants have still to conquer their prejudice against coal." *

While on this subject, I may refer to another, and a still longer, letter of the founder of American liberty, in which he enters in the greatest detail on the inconveniences of every kind experienced by the people who have had neither the skill nor the inclination to supply themselves with coal, or, in default of it, with turf, as they do

* *Lettres de Benjamin Franklin, tome* ii, *page* 42, *des Expériences sur divers Objets de Physique, par M. Ingenhouse; Paris.*

in Holland, where there is a scarcity of wood. "Roads and canals," says he, "by which, in such countries, combustible substances may be cheaply transported from a distance, are of the highest utility, and those who assist in constructing them ought to be ranked among the benefactors of mankind." *

This digression on coal and its useful employment will, perhaps, appear long and tiresome to some readers; but I trust it will be excused on account of the motive which has inspired me to discuss a subject so intimately connected with the comfort of the poor inhabitants of the towns and the country. Unfortunately, most governments are deaf to their own interests: we must not therefore weary of telling them the same thing a hundred times over, when we only have in view the happiness of mankind, and only speak from the evidence of striking examples before our very eyes.

I should have wished to be able to remain at least a fortnight at Newcastle, in order to examine its various manufactures in more detail; but I could not devote more than five or six days to this purpose; for my

* *Idem*, vol. ii. page 419.

principal object being the journey to the island of Staffa, it was necessary not to let the favourable season pass away. We, therefore, made preparations for again taking the road to Edinburgh, and left with regret the town of Newcastle, which afforded us so rich a field of observation. The evening before we set out, we went to take leave of Mr David Crawford, who had shewn us so much kindness, and who was so good as to offer me some specimens of the natural history of the country, which he had selected for me.

CHAPTER VIII

Departure from Newcastle.—Itinerary.—Basaltic Lavas.—Traps.—Porphyries.—Fine Rock of Trap at Dodmill, near Thirleston.—Traps of different colours near Channelkirk Inn

As I had to arrange my notes, and to pack up the different specimens of stones, minerals and coals which I had collected, it was two in the afternoon before we could leave Newcastle; but there was left still time enough of the day to let us reach Wooler, which is at no great distance, and where we slept.

Travellers who love natural history commonly take a pleasure in examining those heaps of broken stones, gathered from right and left on either side of the public roads, which they are meant to repair. They offer an easy means of acquiring, if not a perfect knowledge, at least a fairly correct estimate, of the physical and geological features of a country. In this respect the engineers of bridges and highways [*Ponts et Chaussées*], who have opened up so many communications, and have proved so serviceable to the public, merit also the gratitude of naturalists.

It will have been observed from the Itinerary which I have sketched, that limestone, either in mass, in banks, or in layers, is found all the way until very near Newcastle; but it ought to be remarked, that in proportion as the calcareous substances disappear, their place is occupied by vast deposits of rounded pebbles, which form entire hills, and descend to a great depth in the earth. Sometimes the pebbles alternate with beds of sandstone, gravel, clay, and other transported substances, which cover the coal-mines of Newcastle. This indicates a sudden and rapid revolution, which has produced great displacements, and has gathered together by the action of currents substances frequently of an heterogeneous kind. The line of separation between limestone and granitic rocks is always distinguished by a sort of intermediate zone, of greater or less breadth, which merits the attention of naturalists.

It is there that in general are found coal, hæmatite, ochreous iron, and sometimes lead. With regard to the ores, indeed, there may be some exceptions to this arrangement; but in the numerous observations which I have made, I have found no variation among the other substances. I have always found pebbles, rounded cobble-stones, breccias, conglomerates, sands, and quartzose sandstones, frequently mixed with scales of mica, in this intermediate band, which seems to separate the calcareous hills from the granitic chains.

All the way from Newcastle to Wooler, the heaps of stones collected for repairing the road are composed of black basaltic lava broken in pieces. I do not know whence these lavas are brought: they may have been transported by sea, or by the canals which afford the greatest facilities for bringing them to this place. I saw nothing of a volcanic nature in the place.*

In approaching Wooler, we enter among porphyries, and large blocks of them may be every where seen, lying here and there

* [The stones here referred to doubtless belonged to the district. The numerous exposures of the "Whin Sill" and the great doleritic dykes have long been quarried for road-metal.]

in the fields, swept along by some revolution.* The felspar of these porphyries, less durable than the rest of the stone, has been partly destroyed in some blocks, and appears corroded and carious in others; so that at a certain distance the porosity of these porphyries gives them the look of burnt stones, but they are entire, and have not been touched by fire.† They resemble very much the porphyries of the mountain of Esterelle, in Provence, on the road from Frejus to Antibes, which are equally dotted all over with pores, solely in consequence of the decomposition of the felspar.

From Wooler we proceeded to Cornhill, crossed the Tweed, and entered Scotland,‡ passing by Coldstream, Greenlaw, and Tibby's Inn to Thirlestane.

The country, near Cornhill, is every where

* [These blocks come from the neighbouring Cheviot Hills, which are mainly built up of such "porphyries." The fragments of the Cheviot rocks have been carried far southward by the ice of the Glacial Period.]

† [Had the author been able to explore this region he would have found abundant evidence of cellular lavas and of former volcanic activity, as he afterwards did among rocks of the same age in Scotland.]

‡ [In the original these two clauses are printed together as the name of a place on the route, thus "*Cros-the-tweed-and-denter-Scotland.*"]

sprinkled over with rolled blocks of trap, which very much resembles basalt. It is necessary to pay particular attention to this subject, since these traps, which are absolutely foreign to volcanos, are, notwithstanding, accidentally mingled with real lavas of a black compact kind, which the same revolution has huddled together.* The same order of substances continues almost to Tibby's Inn.

But it is at a short distance from Thirlestane, near Doddmill, and by the side of a bridge, under which flows the streamlet that turns the mill, that the observer must place himself, who is curious to see a deposit of trap so considerable, that it would be difficult to find any where else so many vast masses, and so many varieties laid open.

This local circumstance is owing to the little stream which falls in a cascade over the

* [In Faujas' time it was the creed of the dominant Neptunian school of geology that the so-called "trap" rocks were aqueous deposits and had no connection with anything volcanic. He himself, having minutely studied volcanic rocks in various parts of the Continent, might have been expected to withstand the influence of this Neptunian misconception and to have recognised that these rocks, whether erupted at the surface or injected underneath it, are of truly igneous origin.]

banks of this stone. We can see from the width of its bed, and the ravages which it has committed, that this rivulet, though inconsiderable in summer, must be a furious torrent in seasons of rains and storms. It has completely uncovered the structure of this mass of trap, the different beds of which, more or less thick, are arranged in the form of a staircase, and fully warrant the application of the name which the Swedes have given to this stone.* I certainly did not expect to find, near the road, so fine an object for study. The moment we observed it, we all made for the bed of the torrent, where we remained a very long time examining the different substances, observing their position, their forms, and their accidental peculiarities, and collecting such specimens as chiefly interested us.

Throughout a considerable extent we distinguished parallel beds of black trap, several of which were two feet thick, others one foot, and some four or five inches, and even less. The hardest strata often lie upon a finer-grained, less compact, and less coherent kind, which by the action of the stream has

* *Trappa*, in the Swedish language, signifies steps or stairs.

been destroyed along its length, whereby its continuity has been interrupted and hollows have been formed, on which the feet may be placed, so that one can climb this kind of trap-amphitheatre by the aid of these natural steps.

To the trap, which is hardest, heaviest, and of roughest grain, there succeeds a stone, which resembles it at first sight, but which has a much finer make, and which under the influence of the weather breaks up into very thin leaves, and contains a little mica. It is known that there are traps which decompose in the air, and which, in that state, might be taken for clays. At the same place, an interval may be noticed where the beds are suddenly interrupted by a space in which the prismatic structure has been developed. The material must have adopted this structure at the time of the drying of the masses, accumulated and deposited by the aqueous fluid, for it is certain that there is nothing here of a volcanic nature.*

The trap seems also sometimes to shoot into veins, and into a kind of current, with

* [See editorial notes on chap. xvii. vol. ii.]

holes as large as a man's head; but it appeared that these cavities, the edges and internal parts of which are worn and as it were polished, are due to the action of the torrent, which, with the help of loose stones which it sweeps along, attacks and destroys the more tender parts of the trap.

Finally, it may be remarked, that the mass of trap which I have described is, as it were, implanted in a hill of porphyry, to which it adheres. This last stone is in a very advanced state of decomposition, for it is almost entirely converted into earthy matter.

The hill of Doddmill is contiguous to a ridge of other hills nearly similar, which skirt the highway, and which stretch beyond Channelkirk Inn. Several veins of trap may there be seen to traverse sometimes a rock of porphyry, sometimes a substance of an argillaceous appearance, grey, sometimes greenish, brown, or of the colour of iron-rust, which crumbles, exfoliates, and seems to have the same base as that of the porphyry. The crystals of felspar, which constitute porphyries, are, indeed, in general wanting; but I have collected specimens, in which some of them are to be found.*

The following are the principal varieties of trap which I recognised at Doddmill, or in the contiguous hills on the road to Channelkirk Inn.

1. Hard black trap,† of a fine dry grain, having a resemblance to volcanic basalt, but not, like it, magnetic, is less hard, and yields a powder of a greyer colour.

2. The same trap,‡ intersected by some veins of white quartz.

3. Trap of a blueish-black colour,§ spotted with little specks of greenish and red felspar, which assume no regular crystalline form. These specimens may be considered as exhibiting a passage from trap to porphyry, but which is here only in its commencement.

* A similar substance is found at Renaison, in the mountains of the Forest, where may be seen very fine specimens, in which the porphyric basis is entirely destitute of the crystals of felspar on one side, while on the other they appear in parallelopipeds, and form a perfect porphyry.

† *Corneus trapezius colore nigrescente, vel obscuro.*—Waller, vol. i. page 363. *Trapezum nigrum particulis impalpabilibus, lapis lydius.* Deborn, page 161.

‡ Faujas, *Essai sur les Roches de Trapp.* Variété 16, page 107.

§ *Trapezum spato scintillante rubescente mixtum.* Deborn, page 151.

4. Black trap, of a very fine grain,* tender and fissile; in which some spots of mica may be observed. This variety adheres to the hardest trap: and is sometimes found interposed between beds of solid trap, sometimes between deposits of a substance with a base of porphyry.

5. Brown trap, of a less fine grain,† in which may be distinguished ferruginous spots changed into brown ochre: the colour of this trap is evidently owing to the decomposition of the iron.

These five varieties abound at Doddmill; they are not affected by acids, and have no influence upon the magnetic needle.

6. Violet-coloured trap,‡ adhering to black trap. I only mention this variety here to shew, that the iron, in decomposing, may undergo various modifications in colour; for this violet trap is incontestibly the same as the black trap, with which it forms one mass, and on the face of which may be traced the gradations of change in the colouring principle.

* *Saxum corneo et mica mixtum, saxum corneo micaceum fissile colore nigrescente.* Waller, vol. i. page 420.

† Faujas, *Essai sur les Roches de Trapp.* Variété 6, page 92.

‡ *Op. et p. cit.*

7. Trap,* the ferruginous particles of which are changed into an ochrey red: this modification of the iron has impaired the cohesion of the parts, so that this trap is not so hard as that in which the decomposition of the iron is less advanced.

8. Trap, of a yellowish-grey colour.† If this kind be not examined with attention, it may be mistaken for a sandstone, though it is really of a very different nature: its appearance will easily deceive. Like the other kinds it adheres to black trap, of which it is only a modification, the constituent elements of both being the same.

9. Dull violet-coloured trap,‡ approaching to true porphyry, in consequence of the addition of crystals of white felspar. This porphyritic trap joins on to the black trap, and lies like it in projecting strata, resembling stairs. Some of these strata succeed each other without interruption; while in other cases, they are separated by beds of pure and intact black trap. This porphyry also varies much in colour, in a greater or less degree of

* *Corneus trapezius rubens.* Waller, vol. i. page 362.

† *Corneus trapezius solidus griseus.* Ibid.

‡ Faujas, *Essai sur les Roches de Trapp.* Table synoptique, variété 29, page 148.

hardness, and in its felspar, which appears sometimes in specks and shapeless grains, sometimes in regular crystals. On the one hand, some portions of the material are destitute of felspar, while on the other the contiguous parts contain it.

In short, this vast deposit, this immense accumulation of materials proper for the composition of the porphyries, and which forms a series of small hills, from Doddmill to Channelkirk Inn, seems to exhibit the effect of a sudden operation of nature, of a confused and tumultuous precipitation, which has prevented similar substances from arranging themselves, mutually attracting each other, and obeying those laws of affinity by which the more regular compounds are formed.

Nature, considered under this point of view, is certainly not without interest, for him who delights in studying her magnificent operations. I should like to see this high chemistry sometimes associated with that of our laboratories.

I have, perhaps, dwelt too long upon details which cannot interest every one; but the hills of Doddmill and Channelkirk Inn being situated on the road to Edinburgh, I have thought that I ought to indicate them to those who find among the materials that enter into the formation of mountains, objects worthy of their serious attention. It would be difficult to find elsewhere a place more favourable for this study, since here it may be said, Nature displays herself uncovered, and affords the observer an opportunity of tracing the manner in which she sketches out or perfectly forms porphyry, with a basis of trap.*

The stay we made at the foot of these hills being very agreeable, the time passed quickly away. We did not therefore reach Edinburgh until half past nine in the evening. Our postilions conducted us to Dunn's Hotel, a magnificent inn, decorated with columns; but the inside of which, though very clean, did not correspond with the external grandeur of the edifice.

* [The rocks here described as "trap" are really hardened sandstones, grits, greywackes, and shales—unquestionable sedimentary materials, well-stratified and belonging to the Silurian series of the Southern Uplands of Scotland. That the author should have confounded them in the same series of rocks in which he placed the "toadstones" of Derbyshire is a significant example of the want of any truly scientific and accurate basis for the classification of rocks in the later part of the eighteenth century.]

Next day we called on Doctor Black * and other learned men, for whom we had letters: we took a rapid view of the town, and, notwithstanding the kind reception which the gentlemen to whom we were recommended would have been pleased to give us, we delayed the pleasure of seeing them more particularly, until our return from the Hebrides. The season was far advanced, and already we were in some fear as to the dangers of the sea. We meant, besides, to halt for four or five days in Glasgow, because we should not be able to see that city on our return, having the intention of coming back by Perth: we therefore agreed to make, at present, but a short stay in Edinburgh.

* [Joseph Black (1728-1799) one of the most distinguished chemists of the age, who greatly helped to lay the foundations of modern chemistry and thermometry.]

CHAPTER VIII

Doctor Swediaur.—Prestonpans, its Manufactories and excellent Oysters.—Great Iron Foundry of Carron.—Stirling.—Departure from Edinburgh

I HAD the unexpected pleasure of accidentally meeting, in one of the streets of Edinburgh, a learned German, whom I had seen some years before in Paris, at the houses of the Abbe Fontanna of Florence, and Doctor Ingenhousz, with whom he was connected by scientific pursuits. It was Doctor Swediaur, a physician, living in London, particularly conversant in the knowledge and treatment of venereal diseases, and who had published upon that subject a work full of new views derived from profound study and a skilful practice.*

* Doctor Swediaur is at present in Paris, employed in preparing a second edition of his work, enlarged by another volume. This book will be found to contain some new and curious historical researches, and a number of observations calculated to promote the knowledge of this disease and its treatment. He is to publish it in French.

He told me, that wishing to enjoy a little repose, and to amuse himself with the chemical arts, in which he was skilled, he had quitted the capital of England, and had purchased an estate about five miles from Edinburgh, in the village of Prestonpans, by the sea-coast; where he intended to establish a manufacture of sea-salt, principally with a view to separate the mineral alkali and the muriatic acid.

He invited me to see the works which he had begun to construct, and as I had but a short time to remain at Edinburgh, it was agreed that I should dine at his house next day; which I did.

Prestonpans is very advantageously situated for the establishing of manufactures; its proximity to the sea, and the abundance of coal in the neighbouring mines make it extremely convenient for this purpose. The coal of the place, which is the same as that used at Edinburgh, has the deserved reputation of being of an excellent quality. It burns with a vivid, bright, and long flame; its cinder is grey and light. All that can be said against it is the trifling objection that it burns away rather quicker than the Newcastle coal; but I should prefer for domestic use the Edinburgh coal to that of Newcastle; I do not know any that makes a more agreeable fire.

Swediaur shewed me at Prestonpans the seat of the greatest manufactory of oil of vitriol in Britain. I say the seat only, because the whole of the place is surrounded with a very high wall, which does not let even the chimney tops of the works be seen. A small harbour has been contrived to admit the vessels which bring the sulphur; but every thing is so carefully wrapped in mystery, that the harbour itself is surrounded with walls of a great height. Hence all is concealed in this manufactory, and none can enter but the persons in employment. The only thing known is, that the oil of vitriol [sulphuric acid] which it produces, forms an article of very extensive commerce. I hardly suppose that the processes employed here can differ much from those generally known, which consist in burning the sulphur in chambers lined with lead. The suffocating smell perceived at a distance seems to announce that they are the same. But they may have some processes here for rectification, or other purposes, which they are desirous of concealing.

A great deal of sea-salt is also made at Prestonpans, for home consumption, and as an article of commerce. It is produced by means of fire and evaporation. We found no difficulty of admission to the salt-works, which are pretty numerous.

The sea-water is raised by pumps into immense boilers, or rather a kind of great reservoirs, of an oblong square form, which are not at most above eighteen inches deep, and are constructed of strong sheets of iron closely joined to each other. The reservoir is supported on strong bars of cast-iron. The furnaces are placed immediately underneath, and divide into several vents which reach to the extremities of the boilers or reservoirs. There are four or five of these furnaces to each boiler, according to its surface, and they are supplied with fuel of pit-coal. The water is by this means kept in continual ebullition; and fresh supplies are pumped in, in proportion to what evaporates, until the salt is formed in a quantity sufficiently large to be taken out. By this simple process, there is procured a white salt of very good quality, excellent for cookery and other uses, but little suited for curing provisions, nor so good as French salt for that purpose.

I observed in these salt-works, where artificial ebullition supplies the place of natural evaporation, that the atmosphere is always a little loaded with marine acid in the form of vapour, which quickly corrodes and destroys the polish of steel. I experienced its effects on the buttons of my coat, which were covered with rust in ten minutes. This vapour also affects the smell, and is somewhat injurious to the lungs.

This is certainly not the marine acid disengaging itself from the mineral alkali; their union is too intimate for that supposition. The most violent fire acting upon sea-salt volatilizes rather than decomposes it; an intermediate substance is always necessary for the latter purpose. But there is sometimes found in salt a small portion of the muriatic acid, united with magnesian earth; and as this base fixes it but feebly, it may be disengaged by ebullition.

Doctor Swediaur conducted me to the piece of ground which he had purchased; where the works for making salt were considerably advanced; the boilers being already in position. I saw all this work with much interest.

I ate some excellent oysters at the table of this learned man, as was not to be wondered at, seeing that I was in the place where the most famous oysters are taken in abundance; the rocks at the surface of the sea around the coast are covered with them. They are large, plump, and of an exquisite taste; and are held in such estimation, that they are exported to the principal cities of England and Holland. Large quantities also are pickled, put into barrels, and sent wherever there is a demand for them.

The position of Prestonpans, and its proximity to the city of Edinburgh, make it very agreeable; one who loves study and quiet may here spend some happy hours. I am not surprised, therefore, that Swediaur, fatigued with the bustle of London, should have selected this spot in which to settle, for the more uninterrupted prosecution of his studies and useful occupations.

I spent a very instructive day with him, and returned in the evening to Edinburgh. He had the goodness to accompany me back, with the intention of taking me the next day to Carron, to visit the greatest iron-foundry in Europe, but where it is impossible to obtain admission without very strong recommendations. Swediaur was acquainted there; and I thought myself exceedingly lucky to be able to make this useful excursion under his auspices.

From Edinburgh to Carron is reckoned to be thirty-six miles; but the road is excellent. Count Andreani, Thornton and myself set out in company with Swediaur at six in the morning. We did not alight till we reached Linlithgow; where we took some refreshment. We then proceeded to Falkirk, and at three o'clock in the afternoon reached Carron. The ground from Edinburgh to near Carron is strewn with large round blocks of basalt. The road is metalled with this volcanic lava, broken into small pieces; and there can be no better nor more durable roads than such as are made of this material.

Immediately on our arrival, Swediaur wrote a note to a person belonging to the manufactory, with whom he was acquainted. An answer was returned, that it was necessary to leave the name, designation, and residence of each of us. The demand was instantly complied with; and a few minutes after we were told that we were at liberty to enter.

A man attended us at the gate, who said that he was ordered to conduct us every where, with the exception of the place where the cannons are bored, which nobody was allowed to see.

He conducted us at first into an immense yard, surrounded with high walls and vast sheds. This place was covered with cannons, mortars, bombs, balls, and those large pieces, short and expanded at the breech, which bear the name of *carronades*. Amidst these machines of war, these terrible instruments of death, gigantic cranes, capstans of every kind, levers, and assemblages of pullies, serving to move so many heavy loads, are erected in situations convenient for that purpose. Their various movements, the shrill creaking of pullies, the continued noise of hammers, the activity of those arms which give the impulse to so many machines;—every thing here presents a spectacle as new as it is interesting.

Under the sheds where the finished articles are deposited, we saw several rows of rampart cannon, seige-guns, and field-pieces, destined for Russia and the Emperor. They were longer than usual, of the most perfect workmanship, and covered with a thin varnish, of a steel colour, to preserve them from rust. Their carriages of cast-iron are remarkably simple; they appeared to me to unite the merit of the greatest solidity, to that of being free from the numerous appendages belonging to wooden carriages, which serve only to embarrass manœuvres, to obstruct the march, and to require frequent repairs.

The coating with which the cannons are covered is kept a great secret; but I am inclined to think, that it is composed of a fat, quickly-drying oil, to which is added a certain portion of amber-varnish mixed with some plumbago.*

The large work shops where the cannons are bored are not far from the first yard. We passed close by them; but were very politely told, that particular processes and machines unknown to every other establishment of the kind, rendered it necessary to

* For my own satisfaction, I have made several experiments with these ingredients, and they seemed to answer the same purpose.

keep that place concealed from strangers. We thought this was very reasonable, and followed our conductor to another quarter.*

We were conducted to the works for melting the ore; where four furnaces, of forty-five feet in height, devoured both night and day enormous masses of coals and metal. One may from this judge of the quantity of air necessary to feed these burning gulfs, which disgorged, every six hours, streams of liquid iron. Each furnace is supplied by four air-pumps, of the largest size; where the air, compressed into iron cylinders, uniting into one pipe, and directed towards the flame, produces a sharp whistling noise, and so violent a tremor, that one not previously informed of it, could hardly avoid a feeling of terror. These air-machines, or species of gigantic bellows, are put in motion by the action of water. Such a torrent of air is indispensably necessary to maintain, in the highest state of ignition, a column of coal

* I have seen the fine boring-houses for cannon at the foundry of Creuzot, near Montcenis, in Burgundy. Nothing can exceed the precision of these vast and superb machines, which are moved with water, raised by steam engines: I doubt whether it is surpassed by the engines at the foundry of Carron.

and ore forty-five feet high; and it is so rapid and active, that it projects a vivid and brisk flame more than ten feet above the top of the furnace.

An open area, of very great extent, built in the form of a terrace, and on a level with the upper aperture of the fire-places, is appropriated to the reception of the supplies of ore and coals; and on this platform are also spacious places, where the coal is prepared for use. As the coal employed here consists almost wholly of large lumps, the process by which they convert it into *coke* is completely different from that employed at Newcastle, where only coal-dust is used. At Carron foundry, this operation takes place in the open air, and is of the greatest simplicity. A quantity of coal is placed on the ground, in a round heap, of from twelve to fifteen feet in diameter, and about two feet in height. As many as possible of the large pieces are set on end, to form passages for the air; above them are thrown the smaller pieces and coal-dust, and in the midst of this circular heap a vacancy is left about a foot wide, where a few faggots are placed to kindle it. Four or five apertures of this kind are formed round the ring, particularly on the side exposed to the wind. There is seldom, indeed, occasion to light it with wood; for these works being always in action, they generally use a few shovels of coal already burning, which acts more rapidly than wood, and soon kindles the surrounding pile.

As the fire spreads, the mass increases in bulk, swells up, becomes spongy and light, cakes into one body, and at length loses its bitumen, and emits no more smoke. It then acquires a red, uniform colour, inclining a little to white; in which state it begins to crack and split open, and to get distorted into the shape of a mushroom.

At this moment, the heap must be quickly covered with ashes, of which there is always a sufficient provision around the numerous fires where the coal is prepared in this manner.

The spreading of a tolerably large quantity of ashes over the fire to deprive it of air, is similar to the process used in making wood-charcoal, which is covered over with earth. The result is also nearly the same; for the coal, thus prepared, is light and sonorous, and produces the same effect in high furnaces as wood-charcoal. This is a quality of extreme importance; since, by the use of coal, foundries may easily be established in places where the want of wood would otherwise render it necessary to abandon the richest mines of iron.

There is here such a great series of these places for making coke, to supply so vast a consumption, that the air is heated for some distance, and at night everything is resplendent from the fires and the light. When one observes, some way off, so many masses of burning coal on one side, and on the other side, so many sheaves of flame, darting to a great height above the high furnaces—and at the same time hears the noise of the heavy hammers as they strike on resounding anvils, mingled with the sharp whistling of the blast-pumps—one doubts whether he is not at the foot of a volcano in eruption, or whether he has been transported by some magic spell to the brink of the cavern, where Vulcan and his Cyclops are busy forging thunderbolts.

I wished that Volaire, the painter of Vesuvius, who has so well expressed the terrific aspect of that volcano during its most violent nocturnal eruptions, would come here to exercise his brush on this artificial volcano, which he would find no less striking than the other.

The supplies of ore are on the same terrace with the coals. A canal,* dug at a great expense, and which communicates with the sea, serves to convey all the materials used, and to transport its manufactured productions.

Three kinds of ore are employed here, which are stored up in distinct and separate heaps.

The first consists of a decomposed hæmatite, which is procured from the county of Cumberland. It is reddish, soft to the touch, and stains the hand nearly of a blood colour: it is very rich in iron.

The second is a hard rocky iron-ore, of a yellowish brown colour.

The third, of a deep iron-grey colour, sometimes inclining a little to violet, is remarkable for being formed in geodes [septarian nodules], of a round or oval form, a little flattened. The largest of these geodes are

* The Forth and Clyde Navigation.

about eighteen inches in diameter, and the least from four to five inches.

On placing them on one edge, and giving the other a hard smart blow with a hammer, which breaks them into two pieces, one is agreeably surprised to see their interior filled with a multitude of very distinct, small prisms of three, four, and five sides, and separated from each other by filaments, or streaks, of white calcspar, and sometimes of heavy spar, and white or yellowish spathose iron.

These prisms consist of the same matter with the geodes, that is, of a kind of hard limonitic iron-ore, which has rather the appearance of a dark grey argillaceous stone, than of an ore of iron. The prisms, which can only be considered as the result of contraction, when the substance of the geodes was soft, are from three to four lines broad, and from two to three inches long. They are of a well-pronounced form; and in some of the largest geodes the prisms are so many, and disposed with such order, that they resemble, in miniature, those fine basaltic colonnades, commonly known by the name of giant's causeways.

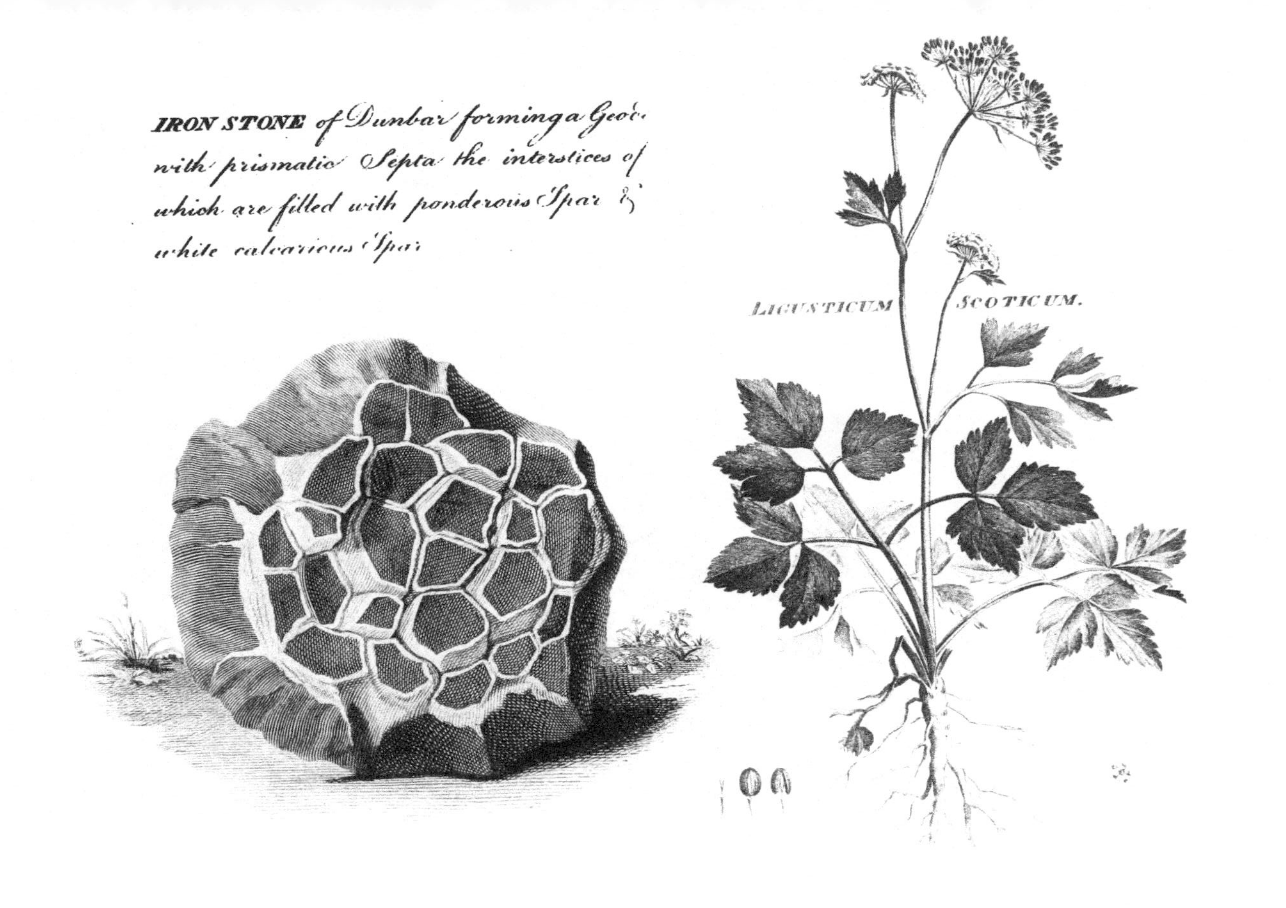
IRON STONE of Dunbar forming a Geo
with prismatic Septa the interstices of
which are filled with ponderous Spar &
white calcarious Spar
LIGUSTICUM SCOTICUM.

This species of iron ore is obtained in great abundance from a hill near Dunbar, a small town in Haddingtonshire, about thirty-six miles from Edinburgh, on the sea-coast, and, consequently, very convenient for transport. It yields a good deal of iron, which one would not expect from its appearance to the eye; but analysis and experience have proved that it is valuable. It must be roasted before being used.

By the due intermixture of these three ores, there is obtained a grey, crude iron, of good quality, which is applied to the most varied uses. It is so soft as to be easily filed; and, as it is pure, it takes the most delicate shapes.

We may readily believe, that it was only by repeated gropings, experiments, and expenses, often fruitless, that this establishment has, at length, reached its present high state of perfection, where every thing is arranged and carried on, with exact precision, and nothing is left to mere routine or chance.

The ore is methodically mixed, carefully weighed, and put into baskets of equal dimensions. The same attention is given to the coal. Every thing is placed in regular order, within reach of the foundrymen, under the sheds appropriated to that service. The baskets for each charge are always counted; a time-piece, which strikes the hour beside the large furnaces, determines the precise moment for putting in the charge. It is the same with respect to the outflow of the melted iron; the clock announces when they should proceed to that operation; and each workman then flies to his post.

We visited the workshops where the crude iron is refined in reverberatory furnaces, to be afterwards cast into cannons, mortars, howitzers, bombs, balls, &c. We saw, also, those where the moulds are prepared, and others where they are dried.

We were then conducted into a vast workshop, which suggested the most pleasing ideas, seeing that the various implements of agriculture, the arts, and domestic use are made there. In the same place they cast boilers five feet in diameter, for the making of sugar in the West-Indies; stoves, in the shape of an antique urn, mounted upon pedestals; grates of all kinds, and in the best taste, for coal fires; kitchen ranges, with all their appurtenances, kettles, tea-pots, sauce-pans, frying-pans, neatly and solidly tinned; spades; hoes of different sorts, for cultivating the sugar-cane, which were ground to a sharp edge on large grind-stones; bas-reliefs after excellent models, for the backs of fire-places; in a word, every thing down to cast-iron hinges and bolts for doors; and most of these last-mentioned articles are so moderate in price, that a man of very limited means may here procure many articles of necessity, and even of ornament, which he could not obtain elsewhere at three times the price. But labour and workmanship are here supplemented by machines and ingenious processes, whereby the work is hastened and made more perfect.

I must not forget to notice a very simple machine, which serves to grind and reduce to a very fine powder the wood-charcoal used for sprinkling over the moulds; it consists of a kind of cast-iron mortar, several feet in diameter, closely shut with a wooden cover, which is perforated in the middle, so as to admit the passage of a vertical cylinder, that forms the principal mechanic power of the machine, being turned round on its own axis by a wheel, which is moved by water.

Two iron bars pass horizontally through the bottom of the vertical axis, where they form a kind of cross, which may be raised or lowered at pleasure, by means of several holes, at different distances, in the axis.

This cross divides the area or capacity of the mortar into four portions, two of which are occupied by two iron balls, nearly as large as ordinary bombs, but entirely solid, and with a polished surface. The moment the axis is put in motion, the balls begin to roll round after each other, and thus speedily bruise the charcoal. But as by this means the latter might be compressed only, without being reduced to a fine powder, the two other spokes are furnished with teeth as in a rake, which stir up the charcoal from the bottom of the mortar, and turn it on every side; so that, in a very short space of time, and so to say, without trouble, whole sacks of charcoal may be ground to impalpable powder, without loss of material.

Conceiving that a correct representation of a Dunbar geode, as it appears on being broken, might interest naturalists, I have had one drawn of the natural size, which will be sufficient to give an idea of the others (Plate I.).* The largest differs from this only in having a greater number of prisms. Sibbald, in the work entitled *Scotia Illustrata*, printed nearly two hundred years ago, gives a representation of this ore, which was known even at that period; but the engraving has been made from a mere sketch, which badly expresses the object.

It is time that I should leave the foundry of Carron, and proceed to other objects. I should have liked to say more about it in fewer words; but, while directing my whole attention to so interesting and complicated a manufactory, I was obliged to trust to my memory for retaining the facts; for it may be presumed that I was not at liberty to take notes of them in writing. I was therefore obliged to take up a part of the night in drawing up an account of my observations. I am well aware that there are some things which I have not fully investigated, and

* In an essay on the Theory of the Earth, printed at Edinburgh, in 1785, Doctor Hutton has also given an engraving of a specimen of the prismatic ore of Dunbar. [These "geodes" are septarian nodules of clay-ironstone which abound in the shales of the Carboniferous formation of Scotland.]

others respecting which I have perhaps gone into too much detail. But when one sees rapidly, that is, when one has not time to see well, one cannot describe well. Others may be able to do better, and I request that they will correct such omissions and errors as have escaped me.

As we were at no great distance from Stirling, we went next day to see that little town, formerly the residence of the kings of Scotland. There is still standing a wing of the ancient palace, which exhibits the remains of grandeur, and is now occupied by the governor of the castle. We likewise visited the parliament-house; it is a hundred and twenty feet long, but is now dilapidated. The oaken doors are covered with bas-reliefs, and rather old inscriptions. I had not sufficient time to have a drawing made of them, nor of some other bas-reliefs in stone, let into some old walls in the public place. These sculptures, which seemed to belong to tombs, are in a singular style, somewhat resembling that of the Egyptians; they consist of figures swathed after the manner of mummy-cases.

This town is very old: the Phenicians

traded to Cornwall for tin; and it is therefore not improbable that that travelling and commercial people might have had some emporium in this part of Scotland. Their monuments, if we may judge by those of Malta, resemble those of the ancient Egyptians, at least as respects the tombs. I merely mention this by the way, in order to induce the Antiquarian Society of Edinburgh to verify or refute my conjectures.

I wished much to pay my respects to Lord Kames,* who has carried agricultural improvement to so high a pitch of perfection, on an estate which he possesses near Stirling; but I was told that he was then at London; so that I was deprived of the pleasure of seeing so worthy a man, whose private virtues and love of rural life have given him a high place in the public estimation.

We went round the upper end of the arm of the sea, called the Firth of Forth, which terminates at Stirling, near the mouth of the river Forth, from which it probably derives

* [Henry Home (1696-1782), a judge of the Court of Session, with the title of Lord Kames, was a conspicuous writer on legal, philosophical and historical subjects, and took much interest in agriculture, on which he published a volume in 1776 under the title of "The Gentleman Farmer."]

its name. We travelled through Alloa, Clackmannan and Culross, where excellent coal-seams are actively mined.

The ground is covered with compact lavas, and others formed by eruptions of volcanic mud. The beds of coal, which lie more than a hundred feet beneath the surface, have remained intact and not burnt by the heat of the lavas above them. But what is remarkable, is that these mines so rich in coal, extend to a considerable distance under the bed of the sea, and that the workmen, guarded against a few leaks by steam engines, which raise the water out of the pits, continue their labour in perfect security, and without the least anxiety from the enormous masses of water pressing above their heads.

Thus, while these bold and indefatigable miners, feebly lighted by the funereal glimmering of their lamps, make these deep excavations resound with the strokes of their mattocks, vessels, borne along with propitious breezes, pass in full sail over their heads; and the sailors, enjoying the fine weather, express their happiness in songs. At other times the tempest lowers, the horizon flashes with fire, the thunder roars, the sea rages, all is dismay, and the crew tremble. But the quiet miners, ignorant of what is passing aloft, joyful and happy, sing in chorus their pleasures and their loves, whilst the vessel is broken in pieces, and swallowed up above their heads,—unfortunately, too true a picture of the daily vicissitudes of human life!

From Culross we journeyed to Inverkeithing, where we crossed the Firth of Forth in a ferry-boat in front of Queensferry, and regained the road to Edinburgh.

We intended, on our return from the Hebrides, to make a long enough stay in Edinburgh to let us become more particularly acquainted with that city and its surroundings; I shall therefore reserve what I have to say about them till that period. We made all our preparation for leaving Edinburgh next day; and that we might have nothing to do but to step into our chaises, we settled our reckoning the evening before. The charges demanded of us were at more than double the rate of those which we had paid at the best and dearest inns on the road; though, as generally happens, we were nowise better entertained. The bill presented to us was more than an ell long, and adorned with flowered ornament and vignettes; and, to prove that nothing had been forgotten in it, they had not failed to set down half a sheet of paper, which one of us had called for, to save the trouble of opening his portfolio. *Paper, three English pence; that is to say six French sous; further for having brought the said paper*, six sous, that is twelve sous. [or sixpence in English money]. We pay without saying a word, but we return no more to Dun's hotel, to lodge under columns less heavy than the rapacious hand of the landlord.

CHAPTER IX

Departure from Edinburgh.—Livingston.—Moorhead Craigs—Prisms of Basalt.—Hearsthill.—Ball of Basalt—Compact Lavas.—Peat.—Pit-Coal.—Glasgow.—Natural History

FROM Edinburgh to Livingston is fifteen miles; the road, as well as the adjacent fields, is strewn with fragments and blocks of basalt. Six miles from Livingston, and at a place called Moorhead Craigs, we observed by the roadside a small peak of basalt, having a tendency to divide into prisms, some of which indeed were distinctly formed.

Hearsthill is three miles distant from Moorhead Craigs. The traveller ought not to neglect to notice here on the left side of the road a splendid natural ball of basalt, which is more than five feet in its greatest diameter, for it is somewhat oblong. The crust or outer envelope, which is very hard and fresh, is nearly three inches thick, and contains another solid ball equally sound, and of the same shape. It is remarkable that the space between the solid ball and the hollow cover is well defined, and more than an inch wide. The crust seems to be completely detached on all sides; though it must necessarily touch the core at some points not exposed to view.

The same revolution which removed and transported to this place a boulder of such bulk and weight, has so conveniently broken off a piece of the crust, that one might imagine it to have been purposely detached in order to disclose the interior structure of the mass.

That lucky accident, that contraction of the lava in all directions at the time of cooling, which produced this kind of volcanic geode, is worthy the attention of naturalists. With pleasure, therefore, do I point out the place where I observed it which may be very easily found. It rests on a slightly raised heath near Hearsthill, six paces from the left side of the road from Edinburgh to Glasgow.

This same place presents another object, no less worthy of attention, and which well deserves a minute examination by the naturalists of Edinburgh or Glasgow, who have it more in their power to inform us of those facts and details, of which the full working out would require a rather long sojourn on the spot.

The ground at Hearsthill is an elevated plateau covered with blocks and fragments of compact lavas, which appear to have been carried thither by some revolution of nature.

At a little distance from the basaltic boulder, above mentioned, and in an opposite quarter, that is, to the right of the road, several small hillocks may be observed, covered with thick and mossy grass, which seems to grow out of a black and marshy soil; yet in this place there is neither marsh nor water.

Some of these elevated spots have been dug into open trenches; and one may there see with astonishment, 1st, a stratum, from two and a half to three feet thick, of good peat, which serves for the use of the district; 2dly, thick deposits of clay, intermixed with masses of basalt; 3dly, a coal-pit in full working order, with its adits open to the day, and lying beneath the deposits just referred to.

This is a curious fact in natural history, and worthy of careful investigation. Had I known that I should have met with so remarkable an object of study, I should have certainly made arrangements for stopping some days at Hearsthill, in order to trace with that attention which such a subject requires, the disposition and order of the several substances, and to measure all the layers. But as we had to reach Glasgow that evening, I had time only to take a general view of the place. What I have stated, therefore, is to be considered as no more than a simple indication, an appeal to naturalists to direct all their attention to so interesting a phenomenon.

On our arrival at Glasgow, our first business was to deliver some letters of introduction, which we brought with us from Edinburgh; and we then went to see such things as were most worthy of notice in the city. Natural history is not so much cultivated here as it is at Edinburgh. Commerce, which is here very considerable, appears to absorb every thing. The University and printing-houses, however, have enjoyed a high reputation, and the city has produced various learned men. We were told of a cabinet formed by Mr Anderson in the University; we went to see it, but we found there only a collection of the most common philosophical instruments, and a few minerals which were in general little interesting.

I was greatly astonished, in a climate so cold and so moist as that of Glasgow, to see the greater part of the lower class of women, and even many of those in easy circumstances, going about with bare feet and bare heads, their bodies covered only with a bodice, a petticoat and a cloak of red stuff, which descends to the middle of their legs; their fine long hair hanging down without any other ornament than a simple curved comb to keep back what would otherwise fall over their faces. This garb of the females, quite simple as it is, is not without grace, and since nothing impedes their movements, they have an elegance and agility in their gait so much the more striking, as they are in general tall, well made, and of a charming figure. They have a bright complexion, and very white teeth. It is not to be inferred, because they walk bare-legged, that they are neglectful of cleanliness; for it appears that they wash frequently, and with equal facility, both their feet and their hands. In a word, the women of Glasgow will be always seen with pleasure by the lovers of fair nature. The children and young folks go also barefooted.

The vicinity of the mountains attracts a considerable number of Highlanders to this city. Their antique costume, much resembling that of the Roman soldiers, forms a remarkable contrast with the dress of the women and other inhabitants. I shall speak elsewhere of this extraordinary garb, which goes back to a very remote period.

In the environs of Glasgow a considerable number of mines yield coal of excellent quality. These make manufactures and commerce to prosper; and thereby increase the happiness of the inhabitants.

This coal is found under beds of quartzose sandstone, which, in some mines, are more than one hundred and forty feet thick; it adheres to the sandstone, without any intermediate material. I tried to find impressions of ferns or other plants in these mines; but they are here very rare; it was only after a careful examination of considerable heaps of material brought up from the bottom of the pits that I was able to recognise in the part of the sandstone adhering to the coal, some characteristic pieces of a large fern, which has some resemblance to the tree-fern of America.

There are also found in a kind of coal which the miners, on account of its variegated and changeable appearance, call *parrot-coal* some portions that show woody fibres.* This coal is less bituminous than other kinds, does not blacken the hand so much, takes fire more easily, and burns with a very bright flame; but is not so lasting in the fire.

The sandstone which overlies the coal-seams of Glasgow, is in general a coarse-grained quartzose rock. Not far from the city, and near some coal-pits, a large quarry has been opened in this sandstone. This excavation is very old, and considerable quantities of stone have been dug from it so that the opening is now very large, and

* [The more usual explanation of the miner's term "parrot-coal" refers it to the chattering noise made by this kind of fuel when thrown into the fire. The coal is not usually variegated and changeable in appearance, but on the contrary is for the most part much duller and more stonelike than other coals. It is chiefly used in gas-making.]

nearly eighty feet deep, where the eye has a full view of the interior of the mass of sandstone.

It is disposed in beds nearly horizontal and more or less thick. But as the substance is homogeneous, one cannot be quite sure whether the lines of separation have been produced by successive deposits, or whether they are merely the effects of shrinkage.

The coal begins to appear at the depth of thirty feet from the surface, in inconstant traces, that run in an irregular manner through the midst of the sandstone. Then follow beds of the same stone, without the least vestige of coal. But as the strata descend, the coal re-appears in small veins, from three to four inches thick, without order or continuity. These are again succeeded by sandstone which is unmixed with coal for a thickness of more than forty feet until it reaches the thick and persistent seams of coal.

The quarry of which I am speaking did not admit of my tracing the order of the beds and the disposition of the several materials beyond a depth of about eighty feet. But having descended into one of the neighbouring pits, sunk through the same stone, I had there an opportunity of making the foregoing observations.

This statement may be of service to those who, from motives of public utility, are engaged in studying the theory of coal-mines, and are desirous of applying their knowledge to practice. The best coals known are in general found under sandstone, both in England and in France.

Let us suppose, for instance, that experience had not already ascertained the disposition of the substances in the coal-mines of Glasgow, and that a pit were opened from above the freestone till the miners had reached the small straggling veins of coal;—if they should then follow, by a lateral course, this first indication, they would certainly be misled. If, on the contrary, they were to continue sinking the pit in a vertical line, they would then discover a second indication in the little veins, rather thicker than the first, though running also in an irregular manner. But as, in continuing to dig, they would have descended more than eighty feet, without meeting another indication, they might, at length, become disheartened, and abandon one of the richest mines, when on the very verge of success, by penetrating a few feet farther.

It is my opinion, that if it were possible to procure topographical plans, carefully executed after accurate drawings from nature, taken by competent men of experience, and exhibiting sections of the most important and best known mines, it would infinitely advance this useful art, and would, at the same time, throw light on the natural history of the subterranean world.

The neighbourhood of Glasgow presents a fertile field of observation, in its assemblage of coal, sandstone, limestone, and volcanic productions, within very short distances of each other. The lavas, however, occupy the higher parts of the ground; and they present such interesting varieties, that I spent the greatest part of my time at Glasgow in studying them attentively, and in taking notes of those objects which appeared to me most fitted to enlarge this important branch of natural history.

The volcanic zone, which comes from a distance, seems to be interrupted here. It is by the side of a water-mill, called the Town Mill, or more properly in the channel of the stream which turns it, that in quitting the city the first products of a grand subterranean combustion are to be seen.

As cultivation around a city, however, must have naturally changed the face of the ground, it is well to remember that cleared land, pasturages, and gardens, hardly allow us to see the primitive state of the place. But as there are here deep hollows as well as barren and naked summits, inaccessible to cultivation, it is to these untouched pieces of ground that the traveller should first direct his attention, because they are the most striking and least doubtful; and, because they gradually lead him to escarpments where quarries have been opened in the lavas, for obtaining stones fit to make good pavements. Near them are found various other kinds of lava, and volcanic currents of mud, to the formation of which water must have contributed as well as fire.

I pursued this course myself; and as the observations which I made may serve to direct the steps of naturalists a thousand times better informed than I am, I shall here transcribe my notes, as I wrote them, without any other pretension than that of faithfully pointing out the different objects which appeared to me worthy of some attention.

The first volcanic hill, where I found prisms of a very pure basalt, is situated at the end of a pond, near a bleach-field. The prisms are very large, and, though not perfectly shaped, are tolerably well defined. This basalt is extremely hard and black, of a very fine grain, and with a matrix so completely fused, that neither schorl * nor any other extraneous body can be distinguished in it. It strongly attracts the magnet, and emits sparks on being smartly struck with steel. Its constituent molecules are so intimately united to each other, that time and the severity of the climate have not in the least injured either the faces of the prisms, which still preserve their hardness and colour, or the body of the mass, which remains untouched, and shows no perceptible appearance of decay.

Passing thence to the opposite extremity of

* [This term in the author's time embraced a large series of minerals which have long been distinguished by different names. It is now restricted to a form of tourmaline, which does not occur in any of the igneous rocks here noticed. It is difficult to decide what mineral the author meant to designate. It may have been augite, or perhaps some form of magnetic iron].

the same lake, and towards the rising ground nearest the public road, which is itself only a continuation of the preceding hill, the compact lava is no longer that beautiful basaltic lava so pure and so black which I have just mentioned. It has indeed the same hardness; but it is intermixed with felspar of a greenish-grey colour, and many small crystals of black schorl, in the form of striated needles, many of which are in a state of decomposition.

This lava is strongly magnetic. It is composed of small columns of well-defined, triangular, quadrangular, and pentagonal prisms, many of which are sound and show a compact texture when broken, whilst others have their crusts decayed to a certain depth. This alteration is so much the more remarkable as it is progressive, and shows itself uniformly on the different faces of the prisms. Thus, for example, if one of the decayed triangular prisms be broken across with a smart blow of a hammer, on inspecting the fractured piece, it will be found that the decayed parts form a very exact triangular zone, spreading sometimes several inches into the interior of the prism, in such manner, that on removing all that has undergone alteration, the remaining core will still retain the same triangular form. If the prism be quadrangular or pentagonal the internal solid core will have four or five sides. This kind of regular and symmetrical progression in the alteration of the respective faces of these prisms, deserved, in my opinion, to be mentioned here. It may interest those naturalists whose studies are principally directed to the natural history of volcanic productions and consequently to all the circumstances connected with their decomposition.

On descending to the foot of the escarpment, opposite to the Town Mill, and towards the tract watered by the stream which turns the mill, the observer sees a small quarry of granitic lava, intersected with veins of calcareous spar, mixed with grains of quartz and iron-pyrites. The lava itself also contains some pyritous specks, where it comes in contact with the vein of calcareous spar. This mixture of calcareous spar and pyrites, which I had before remarked in lavas, almost always announces the vicinity of some solfatara, where the sulphuric acid, disengaged by the heat, rises in the form of elastic gas, combined with the aqueous fluid; whence there result decompositions and new aggregations, of which we should never have conjectured the origin, were it not for the striking instances afforded by the Solfatara near Naples, where nature seems to work, as it were, under our eyes.

I was pointing out these instances to my fellow-tourists and observers, when William Thornton, whose eye is as penetrating as his judgment, exclaimed: "What you tell us appears to me so true, that I think I see, at a little distance, some bleached lavas, which may very probably present to us the remains of an ancient solfatara."

We repaired instantly to the spot, and discovered a large area, where the black compact lava is not only altered in colour and hardness, but so entirely bleached and turned into earth, that it might be taken for a white clay. A few schorls have resisted this decomposition, and are found intact in the lava. Here, likewise, are seen all the different shades of the colouring principle, derived from iron, which, in its decomposition, has given rise to reddish tints, and all the various modifications producible by this great agent of nature.

These altered lavas led us to other adjacent lavas, which had suffered less, but which presented other accidents worthy of remark. We found some granitic lavas * in the shape of balls, several of which were two feet in diameter, while others were no larger than a swan's egg. As they have undergone different degrees of alteration, they exfoliate, and separate as if in layers; so that when neatly broken into two, they display a round and sound core, surrounded with a number of thin shells of lava which seem to enclose it. Some of these balls are found in a detached state; but most of them are, as it were, implanted in the body of the lavas.

But what is most remarkable, there are found in the volcanised zone of the environs of Glasgow, prisms of a well-defined granitic lava, of different sizes, with distinct angles, leaning against each other, and in general, of a quadrangular, pentagonal, or hexagonal form. These prisms have suffered

* [Probably intrusive course-grained diabases or allied rocks. None of them are true lavas erupted at the surface. There is no known solfatara in this district.]

a peculiar alteration, which has attacked the aggregation of their component parts, or rather, which has dissolved the bond that constituted their hardness. Thus they have naturally lost their angles which fall into ruin, and it is singular that, in proportion as the angles are defaced, the central part which is more solid takes the form of a ball; so that these round masses seem to emerge from the middle of the prisms. I once saw something of the same kind in the volcanoes of the Vivarais; but the shapes were not so distinctly marked as here.*

These preliminary and local observations have appeared to be necessary, to introduce a particular account of the volcanic specimens which I gathered in the environs of Glasgow. I made an interesting collection of these lavas, as well as of their principal varieties, to be sent to France. But as it was far from impossible that some accident might befall them before reaching their place of destination, I drew up, on the spot, a brief catalogue of

* [This jointed structure and the weathering of the columns or prisms into rounded blocks which exfoliate externally, while their core still remains fairly solid, are well displayed by the coarsely crystalline rocks of the intrusive sheets of Central Scotland. It is no doubt these rocks to which the author refers as "granitic lavas."]

them, arranged according to the numbers which I pasted on each specimen and transcribed in my journal. I shall, at least, preserve, in case of any mischance, the remembrance of a number of objects which strongly interested me; and I may thus put naturalists in the way of prosecuting the same inquiries, and of giving more interest and expansion to their labours.*

VOLCANIC PRODUCTIONS OF THE NEIGHBOURHOOD OF GLASGOW

Basaltic Lavas

No. 1. A triangular prism of basaltic lava, black, hard, and magnetic.

No. 2. Prismatic quadrangular basalt, of which one part is a compact, black, hard, and magnetic lava, and of a homogeneous matrix, in which no extraneous body can be distinguished, whilst the other part of the same prism consists of a compact porphyric lava, blackish in colour, and interspersed with small irregular crystals of black schorl † and

* It will be seen in the sequel, that this apprehension for the fate of my collections was not groundless. I, therefore, allow the minute of the specimens, which I had collected at Glasgow, to keep its place here.

† [See note on page 208.]

reddish felspar. This prism, which is only eight inches in length, by two and a half inches in breadth, is the more curious, as it is a specimen of a lava, the primitive material of which seems to have belonged to a trap with a porphyric base; since one portion of this remarkable prism has the characters of a porphyry, and the other those of trap. I had before remarked in primordinal rocks the passage of trap into porphyry by the addition of felspar; but it requires a very happy concurrence of circumstances to discover the same appearance in a compact lava of prismatic form; and, from this point of view, this beautiful specimen is worthy of notice.

No. 3. Pentagonal prismatic basalt, of the most perfect regularity in its five sides; pure black, hard, and attractable by the magnet.

No. 4. Basalt in tabular form; black, hard, magnetic, and very fine-grained, with a few needles of black schorl.

No. 5. Black compact lava, the colour of which has been impaired by the action of some elastic gas, or rather by that of an aqueous fluid, charged with some principle, which, in attacking the ferruginous particles of this lava, has destroyed their action on the magnetic needle. This specimen is remarkable for its adherence to a thin layer of limestone. I collected it on the escarpment which rises above the stream near the Town Mill.

No. 6. Basaltic lava, which has lost its colour, and become white, preserving, however, some of its hardness, and especially its rough and sharp grain.

No. 7. Another compact basaltic lava, so altered, that its substance is not only very white, but likewise soft to the touch and as tender as clay. Some small prisms, however, have still preserved their form.*

Granitic, or Porphyric, Lavas

No. 8. A triangular prism, composed of a lava, of which the black base is somewhat

* [Unfortunately the author has given such meagre indications of the actual localities from which his specimens were taken, that it is not possible to identify all of them and thus to compare his nomenclature with that of to-day. He had, however, obviously distinguished the heavy black, close-grained basalts from the much more coarsely crystalline "greenstones" or diabases and other varieties. All these rocks were clearly recognised by him as of igneous origin, but he did not discriminate between sheets which had been intruded within the crust of the earth and those which had been erupted at the surface. He seems to have regarded the rocks round Glasgow as examples of the latter or truly volcanic class; but they are all undoubtedly intrusive.]

scaly, and intermixed with grains and irregular lamellar pieces of reddish felspar, and a few small specks of quartz. This lava, which is strongly magnetic, seems to have for its base a black trap composed of scaly particles, or, if the designation be preferred, a hornblende, or massive schorl; its fractured surface is bright and sound, and is susceptible of a fine polish.

No. 9. Another triangular prism, of a deep iron-grey colour, its base containing a multitude of small parallelopipedal crystals of white, brilliant felspar, intermingled with small irregular plates of the same substance, similar in lustre and brilliance. This prism, which is of a regular, well-defined form, has suffered an alteration on its faces to a depth of three lines. But this alteration has nowise defaced it; only the colour is there changed, and the grain can be easily incised while the centre remains quite hard. The alteration of the particles has advanced so equably and uniformly on all faces of the prism, that the colour to which it has given rise is sharply delimited from that of the unaltered part; so that, on looking at the end of the prism one sees a central triangle inclosed within another, of a different colour. The sound part is magnetic; the altered part is not.

No. 10. Granitic lava of a black base mixed with a multitude of small crystals of yellowish felspar. One of the faces of this specimen is covered with a layer of rose-coloured calcareous spar, and with a thin crust of white quartz, which forms a transparent varnish above the calcareous spar.

No. 11. The same lava as No. 10, with a few pyritous specks in the base, and a layer of white calcareous spar on one of its faces.

No. 12. The same lava, with a multitude of needles of black schorl.

No. 13. Granitic lava, which seems to consist only of reddish-white felspar, and black schorl in needles: these two substances appeared to be in equal proportions.

No. 14. The same lava as the above; but with a greater proportion of the black schorl than of the felspar. The schorl is in long needles still brilliant, although they have been somewhat defaced by the action of fire. It attracts the magnet.

No. 15. Compact granitic lava, of a greenish colour, with some crystals of felspar and plates of mica. This lava has the strongest action on the magnetic needle.

No. 16. Lava of the variety, No. 10, with crystals of garnet, of twenty-four trapezoidal faces, greenish-grey in colour, and much resembling that found at Vesuvius. This is the first time that I have seen this kind of garnet anywhere than in the lavas of Vesuvius. I have never met with it in the volcanic products of Etna, of the Isle of Bourbon or of Iceland, nor in those of Auvergne, the Vivarais, the Velay, the banks of the Rhine, etc. I found near Glasgow only two specimens of the lava containing these garnets; M. de Mecies picked up a third. The crystals are in the most perfect preservation.*

No. 17. Prismatic porphyric lava, in which there is abundance of dull white felspar, and black schorl in needles. This schorl has lost its lustre, and the base of the lava is changed into an ochreous, friable, and tender substance, which may be picked to pieces and even cut with ease. But notwithstanding

* [The Vesuvian crystals referred to in the text are those of the mineral Leucite. No example of this mineral has been found in any of the igneous rocks of Central Scotland, and it is uncertain to what mineral the author here refers.]

its advanced state of decomposition, this lava strongly attracts the magnet; this may be owing to the schorl, which is not so much altered.

No. 18. Porphyric lava, in the form of a ball in which the concentric layers have been formed by decomposition. In some of these balls as many as seven layers may be counted, which exfoliate and can be detached; whilst the part which has undergone no alteration, and which forms the core, has a bright, hard matrix, and presents no trace of any layers.

Such are the volcanic products which I collected in the environs of Glasgow, during three days which I devoted to these researches without information or conductor. I should have liked to be able to devote more time to this work, which vividly excited my curiosity; but I have given here more than enough to put others in the way of completing the slight sketch which I have just traced. I ought not to forget to say that I found on the highest hill in the vicinity of Glasgow, among rounded lavas, several blocks of a quartzose rock mixed with mica, and containing brown twelve-sided garnets, which though somewhat coarse internally, were very regular in form. These blocks of micaceous quartz, which are found rolled and dispersed here and there, have been swept along pell-mell with the lavas, by the effect of some revolution. But as they are not small in size, and as it is consequently probable that they have not been transported from a great distance, these stones may afford some indications of the primordial rocks in which the volcanoes burst forth in this part of Scotland.

It would also be nowise surprising to find that they have been torn away from a certain depth within the earth by the effect of volcanic explosions; for Vesuvius affords instances of particular rocks, which it throws out in some of its eruptions, but to which no analogues are known either in the neighbourhood or even at a distance. The quartzose blocks with these intermingled garnets, here mentioned, may have had a similar origin, if there should not be discovered around Glasgow, and even at a great distance from it, any rocks to which these blocks might have belonged.*

* [Faujas here displays his strong volcanic proclivities. The blocks of quartzite, schists, granites, and various other

Here I ought to state a difficulty which I experienced in my examination of those lavas of the environs of Glasgow, which I have termed *Granitic* and *Porphyric* lavas.

These lavas undoubtedly come in some cases from the former, and in others from the latter of these compound rocks; there were instances in which I could make the distinction, without fear of being deceived, when their characteristic differences were apparent. But as the long-continued action of fire, as well as that of divers gaseous emanations, has altered and often destroyed the constituent principles of these lavas, while nevertheless preserving some characters common to granites, and porphyries, such as crystals of felspar or schorl, I must necessarily have laboured under some embarrassment and uncertainty. As the base of ordinary porphyries, however, is composed of those particles which constitute trap, or if the term be preferred, hornstone, and as this base readily fuses, it may be distinguished

crystalline rocks scattered so abundantly over the surface of Central Scotland have had no connection with volcanoes, but have been carried by the ice of the Glacial Period from the north, where their parent rocks form the mountains of the Highlands.]

with a little practice. But when the emanations have acted on this base, and have destroyed its cohesion, and when the same cause has produced the same effect on the base of the granites, the observer is involved in an embarrassing uncertainty. I have probably however, said too much upon a subject which can interest only one class of readers.

Our harvest in natural history being finished in this quarter, we had to make preparations for our departure. We were about to enter a mountainous region, and as no relays could be got on the road to Inverary, we hired horses and drivers to be kept during the rest of our journey.

I forgot to mention that we brought with us a draughtsman from Edinburgh, to take such views as should appear to us the most important for the advancement of the natural history of volcanoes in the part of the Hebrides which we were going to visit.

CHAPTER X

Departure from Glasgow.—Dumbarton.—Volcanic Substances.—Loch Lomond.—Luss.—Tarbet.—Loch Fyne.—Inverary.—Duke of Argyll's Castle; its Parks and Gardens.—Natural History.—Departure from Inverary

ON the afternoon of the 14th of September [1784] we left Glasgow with the intention of sleeping at Dumbarton. When we arrived there, we found a fair going on in the town, which made it difficult to find beds at the inns, as these were full of strangers. Here the traveller must bid farewell to English cleanliness: other manners and other customs now appear; but all that can be borne when one is in quest of instruction. The barley and oats were not yet ripe, so little advanced was the harvest at Dumbarton.

This little town stands upon an arm of the sea into which the Clyde, which flows past Glasgow, falls. It is defended by a a small fort, built upon the top of an isolated volcanic peak, which has two summits. I know not why Mr Pennant, in speaking of the rock on which Dumbarton castle stands, should say that its height is astonishing. I found it to be at the most two hundred and fifty feet.

It is formed of a blackish basaltic lava, which is hard, magnetic, with a fine grain and a bright structure. This lava has had in general, a tendency to take the prismatic form; but, with the exception of some small prisms, which may be found here and there, the principal masses present mere outlines of columns.

The face of the hill opposite to the houses is the only part which should for a little engage the attention of naturalists. We there find: 1. A current of muddy lavas, which has taken up a multitude of fragments of more or less altered basalt: this current is crossed by some veins of calcareous spar, due to infiltration. 2. In the same current we also observe a small quartzose zone, of a dull white colour, mixed with rose-coloured calcareous spar. 3. A much thicker vein of blackish argillaceous shale, which breaks into laminae.

I believe that these last substances are still in their primitive place, and have escaped the action of the lavas which have risen up in the midst of them. We may consider them as proofs that volcanoes have here exerted all their fury upon shaly and somewhat micaceous rocks, crossed in some places by seams of quartz.

In our excursions round Dumbarton, we saw immense masses of basalt reduced into fragments. These hard black lavas thus broken, and heaped one above another, form whole hillocks. It is truly astonishing to see such a great accumulation of lavas in fragments, particularly at a quarter of a league from Dumbarton, on the Glasgow road, where these lavas form a vast causeway, which runs on to join some higher hills in the distance.

We were told that these were remains of a wall, astonishing for its thickness and length, which the Romans were obliged to erect in the time of Agricola, to secure themselves against the incessant incursions of the indomitable Caledonians, whom these conquerors of the world were never able to subdue.

I know that ancient authors have spoken of this famous wall, and have said that it was repaired by order of the emperor Hadrian, upon which account it has been called *Vallum Hadriani*; but it was not here that the wall in question was situated.* It is, however, true that Lollius Urbicus, Hadrian's lieutenant, passed the wall of Agricola, and drove the Caledonians beyond the Clyde, where he formed a chain of fortifications.† Dumbarton being near the Clyde, tradition has placed here the position of the Roman wall, which cannot but be that which was erected by Lollius.

I was not surprised, therefore, that the Romans should have taken advantage of a position so favourable for erecting redoubts, which must have been all the easier to construct, since nature has defrayed the first expense, inasmuch as she has accumulated through the active force of volcanoes, an endless amount of materials which she has aggregated and heaped up till they form of themselves a formidable barrier.

* The best maps represent this line of circumvallation as running from Newcastle to Carlisle.

† Capitolin in Antonin V.

I employed more than two hours in examining and traversing this kind of natural causeway, in which I only saw little hillocks of fragmentary basalt, without being able to discover the smallest trace of anything artificial. It is possible, however, that military works may have existed there, and redoubts built of dry stone by the Romans. It would not be surprising that their character should be effaced, when we know that works of the same kind made elsewhere during the wars of Louis XIV. can now scarcely be recognised.

The whole of my collection, then, was made within a distance of two miles round Dumbarton, and was confined to basaltic lavas, containing some kernels of calcareous spar, and a lava of a muddy nature, to which there was attached a rather fine piece of somewhat greenish zeolite.

There are also not far from the town sandstones, in place, of a red colour, which appear to have been touched by fire, whereby their ferruginous colouring principle has been revealed.

We left Dumbarton at five o'clock in the evening, intending to sleep at Luss, on the banks of Loch Lomond, so as to be able next day to examine at our ease that beautiful lake, and its little islands, about twenty-eight in number, on several of which, we were told, there are charming habitations. This fresh-water lake is the largest in Scotland. It measures twenty-eight miles in length, and is regarded as one of the wonders of the country.

The volcanic substances disappear as one approaches the lake. They are at first succeeded by limestones, afterwards by granitic schists, and by micaceous gneiss. But we had scarcely gone a mile along the banks of the lake when night came on, and the weather became overcast: we saw only a few islands, which appeared most picturesque as well as the surroundings. It was ten o'clock when we arrived at Luss. This place being marked on the map, I expected that it was a village, or at least a hamlet. It was, however, only a single, sorry habitation by the side of the lake; and such a habitation! I thought we were entering a fishing hut. But our astonishment was much greater, when they made signs to us not to speak, so as not to disturb the repose of somebody who was asleep.

We believed there must be some one ill in the house; the expressive gestures of the mistress and three other persons sitting in the little kitchen seemed to signify as much. So we dared not open our mouths, and as they well knew before hand what we were going to ask for, they led us, or rather pushed us, into a kind of stable, where we might have a brief audience; and it was, indeed, not a long one. "The lord Judge," said the hostess, does me the honour to lodge here when on circuit. He is there; everybody must respect him. He is asleep. His horses are in the stable; so you see there is no more room for yours; have the goodness then to go away."—"But mistress," said one of our postillions, for we durst not venture to speak, "look at our poor horses, look what a terrible rain."—"Very well, let us look," said she. We went out, and she added "No noise: don't disturb his lordship's sleep, respect for the law. May you be happy! and be off." And she shut the door and double-locked it after us.

We could not help laughing at this laconic eloquence, which admitted of no reply; such respect for a judge is a fine thing. We left, more concerned for our poor postillions and our horses, than for ourselves.

We had still unhappily fifteen miles before us in a dark night, and frightful weather, always along the banks of the lake, without coming upon the smallest habitation. Never in my life did I make so disagreeable a journey, nor one which appeared so long.

Our horses, though good, were fatigued, and with difficulty carried us forward. Our poor postillions wished all the judges in the world a hundred times to the devil, and poured forth a thousand insults on the landlady of Luss. We tried to console them the best way we could, promising them a recompense, which indeed they justly earned; for they were soaked to the bone by a cold rain. At last, after many anxieties and troubles, we arrived at half past three in the morning at a hostelry, equally lonely, called Tarbet.

The people of the house had the kindness to get up at the call of our postillions, and our horses were put into a stable. There was no judge here, but there were jurymen on their way to Inverary, who, having arrived before us, occupied all the beds. But at last our horses were in shelter. We were courteously received, got a morsel to eat, and some excellent tea to warm us.

The calm way in which we took our adventures interested the landlady; and when she saw that we were going to pass the rest of the night in our carriages, this worthy woman came to offer us two mattresses from her own bed, saying that she had had enough of sleep, and would not go to bed again. We thankfully accepted them. Count Andreani preferred sleeping in his carriage. M. de Mecies kept one of the mattresses; Thornton and I shared the other. We slept three hours wrapped up in our cloaks, and our fatigue disappeared.

The most lovely day succeeded to the ugliest night. The sun was brilliant and warm; the sky a fine azure. We breathed pure air on the banks of the lake, and saluted the nymph who presided over its beautiful waters.

From this point of view the appearance of the lake is superb, though only a part of it can be seen, on account of its great extent. It is interspersed with little islands, several of which are only barren rocks, but others show little cultivated plots, and hillocks picturesquely grouped. Some of considerably larger size we could see in the distance with the help of our glasses.

The banks of the part of the lake near us were surrounded with rocks of mica-schist, the layers of which, in wavy ridges, shine as if silvered. A multitude of mosses, mostly in flower, formed little clumps of verdure in the shelter of these rocks, while the more elevated parts displayed pastures, covered with black cattle in the midst of flocks of white-wooled sheep. The shepherds, seated under the pines, and conspicuous by their clothes in patterns of large checks of various colours, gave life to this rural scene, where everything seemed to breathe calm and gentleness. This beautiful prospect forms a fine contrast with the ordinary aspect of the mountains of Scotland, so severe from the sombre colour of the heather, and from that which is characteristic of the remains of ancient volcanoes in those places where lavas abound. *

We much regretted that we did not find a lodging at Luss, where we might have em-

* [Here again the author's passion for volcanoes leads him to see traces of them in the south-western Highlands where they are not to be found.]

barked on the lake, and visited some of the islands, so as to reach Tarbet by water; but there was now no more time; we had to think of continuing our journey. After a walk of an hour and a half, we returned to our inn, where a breakfast of tea awaited us. It was a kind of coquettishness on the part of our hostess, for she had arranged her china cups on a well-painted and well-varnished tea-table, adorned with every accessory of an elegant breakfast, at least in the country. This good woman, who was a widow, had all the simplicity of manners, sensibility and gratitude, which distinguish the inhabitants of the mountains. She hastened to let us know that this little set of movables was the most precious thing in her house, since it had been given to her by the Duchess of Argyll, who had been so good as to lodge at this inn on her way to Inverary. She gave us a warm panegyric on the goodness and generosity of this lady, and at the same time, extolled to us all the good qualities of this amiable and benevolent family.

It was a very great pleasure to me to see this worthy woman thus opening her heart in gratitude, not so much for the present itself as for the esteem with which she regarded the hand that had given it. She did not cease telling us how much the ducal house is beloved in the country. Encomiums of this kind are rarely suspicious.

Persons of fortune or rank have many opportunities, if they so desire, to make themselves beloved, and others happy, at small cost. Why then does the thing so seldom happen? Why? Because it comes more from character than from education, and from combinations of policy and interest; for the natural inclination comes uppermost among men. From all that was related to us here of the Duke of Argyll and his house, and from the tone in which it was told, I was convinced that this family is naturally good, full of excellent qualities, and that in whatever situation chance or fortune had placed them, they would have displayed the same amiable character. Philosophers have not sufficiently studied the passions from the side of nature. Every thing cannot indeed be ascribed to them, but we must assign to them a great deal.

The superb Loch Lomond, the fine sunlight that gilded its waters, the silvery rocks that skirted its shores, the flowery and verdant mosses, the black oxen, the white sheep, the shepherds beneath the pines, the perfume of the tea poured into cups that had been given by kindness, and received with gratitude, will never be effaced from my memory, and make me cherish the desire not to die before again seeing Tarbet. I shall often dream of Tarbet, even in the midst of lovely Italy with its oranges, its myrtles, its laurels, and its jessamins.

But let us proceed upon our journey. I soon found a contrast to the delightful scenes we left. They were succeeded by deserts and dismal heaths. We entered a narrow pass between two chains of high mountains, which appear to have, at a very remote period, formed only one ridge, but which some terrible revolution has torn asunder throughout its length.

This defile is so narrow, and the mountains are so high and steep, that the rays of the sun can scarcely reach the place and be seen for the space of an hour in the twenty-four. For more than ten miles, which is the length of this pass, there is neither house nor cottage, nor living creature except a few fishes in a small lake, about half way. I do not mention the flocks of sheep that feed on the heights, because they are at so great an elevation, and among heathery slopes so steep, that as neither their shape nor their movements can be distinguished, they may be taken to be stones rather than living animals; but they can be made out when one watches them with glasses.

We travelled thus for nearly six hours in this dismal traverse, through which the roads are neither metalled nor kept in repair, until at last we suddenly debouched on the shore of Loch Fyne, in Argyllshire. The first village met with at the end of this lake is Cairndow. Going round this point of the lake, which forms a kind of fork, we arrived at Inverary, the capital of Argyllshire. It must not be imagined that this chief place is a town: it is merely what would be called a village in France; but a village pleasantly situated upon the side of the beautiful Loch Fyne, which may be navigated by large vessels, and abounds with herring at the proper season. This fishery yields a considerable revenue to the country. On one side, there is the view of pasture-grounds and some trees in the valley, which ends in a fine park; on the other, diversified gardens, meadows covered with flocks, and hills planted with green trees, at the bottom of which a superb and vast habitation, in the Gothic style, gives life to this fine landscape. This is the castle of the Duke of Argyll, about a mile from Inverary.

We were standing at the door of the only inn in the village, whence we were enjoying this fine prospect; our carriages were already within the yard, when the landlord came to tell us, very politely, that we could not be received, as every room in his house was either engaged, or already occupied. It was again the lord-judge who was expected here, and for whom the best room had very properly been reserved: the jurymen were in possession of the rest of the house.

We had letters of recommendation to the Duke of Argyll,* and we knew that he had come to pass the autumn in this beautiful country; but we did not wish to wait upon him until after having procured quarters elsewhere, as we should have been sorry to abuse any kindness that might be shewn us.

* [John, 5th Duke of Argyll, born 1723, died 1806.]

The inflexibility of our host, however, cruelly embarrassed us. He would not let us have our luggage detached, nor set foot in his house. No other hostelry existed in the place. Our only alternative was to push on to Dalmally, about fifteen miles from Inverary; but it was already too late to undertake the journey, and we should have been obliged to travel during a part of the night by very bad roads. Besides, we should thus have lost the opportunity of seeing the Duke of Argyll, delivering our letters, and receiving from him information respecting the country, and the difficult route we had still to pursue through such a desert region before we could arrive at the port of Oban.

These considerations induced us to ask the innkeeper, if he would permit us to step into a room, and write a letter to the Duke of Argyll. This name was held in such esteem that instantly every thing we wished for was granted. We stated our situation to the Duke, informing him at the same time our desire to present our respects to him and expressing our reluctance to give him any trouble on that account. To this billet we joined our letters of recommendation. An express was dispatched with the packet, and a reply was promptly brought to us by a French painter, who was working at the Castle, and who came to tell us, that we were eagerly expected, and to beg us to come just us we were, as the family would not sit down to dinner till we arrived. Servants were, at the same time, sent to take care of our carriages.

On our way, we saw the Duke's son,* who came to meet us with manifestations of the frankest politeness and the most gracious affability. We were received at the house with every mark of friendship in the midst of a numerous company and an amiable family, who, joined to the most polished manners those prepossessing dispositions which are the natural dowry of sensitive and well-born minds. After the first compliments, we placed ourselves at the dinner-table, and as every thing pleased and interested me in this house, which was pervaded, if I may use the expression, with a kindly sympathy, I said to myself—"The good woman of Tarbet was right. Here is indeed

* [This was George William (1766-1839) who afterwards became 6th Duke of Argyll.]

a charming family." French was spoken at this table with as much purity as in the most polished circles of Paris. They did not fail to enquire the object of our journey to a country so little visited as this remote part of Scotland; but they were not at all surprised when they heard that our purpose was to go to the Isle of Staffa, with its Cave of Fingal, which had now a great reputation in the country.

If I recollect aright, they told us, that Sir William Hamilton, Ambassador at Naples, and his nephew Lord Greville, had come here with the same intention, without having found a day favourable for the short passage between the mainland and Staffa. As this cliff-girt island has neither harbour nor anchorage, and can only be reached in very small boats, settled weather and a calm sea are absolutely necessary. These conditions, however, are extremely rare upon this coast, strewn as it is with islands, washed by currents, and exposed to impetuous winds.

To shorten the passage by sea, we were advised to make for Oban and coast thence up the Sound of Mull to the island of that name: to cross the whole breadth of the island to

Torloisk, where we should find the house of Mr M'Lean, a very worthy gentleman, to whom the Duke of Argyll promised us letters of recommendation. From Torloisk to Staffa the crossing and return can be made in one day, by setting off in the morning early, and getting back somewhat late. But even for this little voyage the traveller must have one of those fine days which we had little reason to expect at a season already too far advanced. We were told, however, that the autumn might chance to have some of these fine days; and that moreover, we should not have been better off had we come sooner, seeing that the sea had been stormy for some months.

The Duke of Argyll kindly said, that he wished to have the pleasure of detaining us for at least a few weeks, that we might have an opportunity of becoming well acquainted with the country, and such of the neighbouring mountains as were particularly worthy of note. But, pressed for time, we thought that three days, well employed, would suffice to enable us to see what was most remarkable around the castle of Inverary, particularly some rather high hills and some open quarries, and that if we began our work early in the morning, we might devote part of the evening to the duties of society and to the pleasure of becoming more intimately acquainted with a family so unaffected, so well informed, and so worthy of respect.

We remained, then, three whole days in this delightful retreat, devoting the mornings to natural history, and the evenings to music or conversation. As the gentle and amiable manners of the master and mistress of the house strongly interested me, as well as the friendly tone of his children, who were clever and had a thirst for knowledge; and as, besides, I saw here some customs connected with Scottish frankness and hospitality, I shall give a rapid sketch of my observations and remarks. These naturally ought to precede what I have to say on the natural history of the environs of Inverary.

Inverary castle is entirely built of squared blocks of a grey-coloured *lapis ollaris*, soft to the touch, capable of taking a fine polish, as well as every form that the chisel can give it: though tender, it resists the weather, at least as well as the most durable marble.

One is at first sight surprised that a castle, in appearance so ancient, should show not the slightest mark of decay: every part is so well dressed, the angles are so clean and perfect, and the colour of the stone is so equal, that the building seems to have just come from the hand of the workman.*

My astonishment on this subject, however, soon ceased, when, after crossing some draw-bridges, and passing through a gateway, as Gothic as that of the time of Charlemagne, I arrived at a fine vestibule, which led to a staircase in the Italian style, with double balusters, of the best taste and the most perfect architecture.

This vestibule was ornamented with large bronze vases, of antique shape, placed on their pedestals, between the columns. These vases served, at the same time, as stoves to warm the air of the vestibule and the staircase.

The whole enclosure of this staircase is magnificent, tastefully decorated, and skilfully lighted. The steps are covered with elegant

* [Inverary Castle was built (1744–1761) of stone obtained from quarries in the neighbourhood, and had not been finished more than about twenty-three years at the time of the author's visit.]

carpets; every thing here proclaims a love of the greatest neatness. There appears, however, to have been a desire to recall even here some reminiscences of the Gothic, for in the perspective of the staircase, a large niche, ornamented with groups of Gothic columns, has had placed in it a large organ-case: which gives an imposing and religious air to the place. This contrast may appear somewhat odd in theory, but it has been done with a certain charm, which is not without merit.

The rest of the house is laid out in a manner equally elegant and commodious, and can accommodate a large company. As ought always to be the case in the country, much more attention has here been paid to the luxury of simplicity, and the extreme of neatness, than to the display of gilding, and sumptuous furniture.

Notwithstanding its antique appearance, this castle is quite a modern building. The Gothic style was selected, coupled with the best design for the interior, because buildings of the tenth century look well amidst woods, and at the foot of hills. They recall ideas of chivalry connected with the bravery and gallant adventures of those romantic times. These recollections diffuse a kind of charm over the scene: they embellish it, and make it impressive. We are all a little fond of romance.

The parks, planted with foreign trees side by side with those of the country, are of great extent and have the finest effect. Open spaces, covered with the most beautiful verdure, have been carefully left and are traversed by roads and foot-paths which lead to gardens, green-houses, sheep-folds, and sequestered woods, on the sides of hills, the banks of rivulets, or towards the shore of an arm of the sea.*

There were staying in the castle, at this time, the Duke of Argyll, one of the best of men, who had travelled in Italy and in France; the Duchess, who was first married to the Duke of Hamilton, and after his death, to the Duke of Argyll: she passes, and justly, for having been one of the most

* Knox, who was at Inverary two years after me, in speaking of that place, says: "Inverary has become of some importance, by the care of the family of Argyll, who have a magnificent house there, surrounded with more than a million of trees, which occupy several miles square." *Tour through the Highlands of Scotland and the Hebride Isles*, Vol. I.

beautiful women in Great Britain: she is certainly one of the best informed; * the Countess of Derby, the Duchess's daughter, by her first marriage: this lady had travelled a great deal, and speaks French with so much ease and so little accent, that she might be taken for a native of Paris: there are few women more amiable or more beautiful. The children of the Duke were at home. The eldest daughter sings exceedingly well, and plays admirably on the piano-forte, and she, as well as her younger sister, has the sweetest and most lovely expression. The Duke's son, about sixteen or seventeen [actually eighteen] years of age, has the courtesy and kindness of his father, and already shews much skill in drawing. A physician and chaplain formed the rest of the family circle. There were also several visitors in the house, among whom was a member of Parliament, a man of much in-

* [This was the celebrated Elizabeth Gunning, daughter of an Irish country gentleman, and one of the greatest beauties at the court of George III. She had two sons and a daughter by the Duke of Hamilton. Her family by the Duke of Argyll consisted of three sons and two daughters. The eldest son had died in infancy, and the second, now the heir to the title, was only eighteen years of age at the time of the author's visit.]

telligence, who had travelled, with advantage, in almost every part of Europe.

I must not omit to mention, that on the second day after our arrival, the lord-judge, who had ousted us so often, came to dine at the castle. He was already advanced in years, but a good loyal Scot, worthy of all the respect that had been paid to him, for he filled his office with justice and humanity. We made our peace with him in the midst of toasts; and he assured us, with great good nature, that he would have shared his lodging with us, if he had known what passed, and that we might be assured that we should not sleep in the shed, if he had the pleasure of meeting us another time upon the road.

Such was the quiet and kindly way in which we spent our time in Inverary castle. Compare it with life in towns! Each person rose in the morning at any hour he pleased. Some took a ride, others went to the chase. I started off at sunrise to examine the natural history of the neighbourhood.

At ten o'clock a bell gives warning that it is breakfast-time: we then repair to a large room, ornamented with historical pictures of the family; among which there are some by Battoni, Reynolds and other eminent Italian and English painters. Here we find several tables, covered with tea-kettles, fresh cream, excellent butter, rolls of several kinds, and in the midst of all, bouquets of flowers, newspapers, and books. There are besides, in this room, a billiard-table, pianos, and other musical instruments.

After breakfast, some walk in the parks, others employ themselves in reading or in music, or return to their rooms until half-past four, when the bell makes itself heard to announce that dinner is ready; we all go to the dining-room, where the table is usually laid for twenty-five or thirty covers. When every one is seated, the chaplain, according to custom, makes a short prayer, and blesses the food, which is eaten with pleasure, for the dishes are prepared after the manner of an excellent French cook; every thing is served here as in Paris, except some courses in the English style, for which a certain predilection is preserved; but this makes a variety, and thus gives the epicures of every country an opportunity of pleasing their palates.

I was particularly pleased to see napkins on the table, as well as forks of the same kind as those used in France. I do not like to prick my mouth or my tongue with those little sharp steel tridents which are generally used in England, even in houses where very good dinners are given. I know that this kind of forks, which are sometimes placed in a knife handle, are only intended for seizing and fixing the pieces of meat while they are cut, and that the English knives being very large and rounded at the point, serve the same purpose to which forks in France are applied ; that is, to carry food to the mouth[!]. But, I must confess, that I use the knife very awkwardly in this way. As it is well, however, to take account of the usages of different countries, it seems to me that at table, as well as elsewhere, the English calculate better than we do.

In England, the fork, whether of steel or even of silver, is always held in the left hand and the knife in the right. The fork seizes, the knife cuts, and the pieces may be carried to the mouth with either. The motion is quick and precise. The manœuvres at an English dinner are founded upon the same principle as the Prussian tactics — not a moment is lost.

In France, the first manœuvre is similar to that of the English; but when the meat is cut in pieces, the knife is laid down idle on the right side of the plate, while the fork on the other hand passes from left to right which makes the first loss of time; the right hand seizes it and lifts the morsel to the mouth, thus making a threefold manœuvre. The English plan is better, but it necessitates large blunt knives rounded at the point. Well! what harm is there in that? It would mean one weapon less in the hands of fools or villains.

How many beings in illness or in despair have made use of sharp-pointed knives against themselves? How many monsters have made a cruel use of them against others? The list would, doubtless, be long; and, if this useful instrument had not in Italy, Spain, and most other countries taken the form of a stilletto, it is probable that such crimes would be less frequently committed. Experience has long since proved that great effects may spring from very trivial causes.

But I am forgetting that the knives and

forks at the Duke of Argyll's table are used in eating very excellent things. The entrées, the rôti, the entremets are all served as in France with the same variety and abundance. If the poultry be not so juicy as in Paris, one eats here in compensation hazel-hens, and above all moorfowl, delicious fish, and vegetables, the quality of which maintains the reputation of the Scottish gardeners who grow them.

At the dessert, the scene changes; the cloth, the napkins, and every thing vanish. The mahogany table appears in all its lustre; but it is soon covered with brilliant decanters, filled with the best wines; comfits, in fine porcelain or crystal vases; and fruits of different kinds in elegant baskets. Plates are distributed together with many glasses; and in every object elegance and conveniency seem to rival each other. I was surprised, however, to see on the same table, in so cold a climate, and in the middle of the month of September, beautiful peaches, very good grapes, apricots, prunes, figs, cherries, and raspberries, though the figs could hardly be called juicy by a person born in the south of France. It is probable, however, that the greater part of these fruits were produced with much care and expense in hot-houses.

Towards the end of the dessert, the ladies withdrew to a room destined for the tea-table. I admit that they were left alone a little too long; but the Duke of Argyll informed me, that he had preserved this custom in the country, in order that the people of the district might not be offended by the breach of an ancient practice to which they had always been accustomed. Although the ceremony of toasts lasts for at least three-quarters of an hour, no person is made uncomfortable, and every one drinks as he pleases. This, however, does not prevent a great number of healths being drunk with pleasure and good grace. Wines are the great luxury of the table in England, where they drink the best and dearest that grow in France and Portugal. If the lively champagne should make its diuretic influence felt, the case is foreseen, and in the pretty corners of the room the necessary convenience is to be found. This is applied to with so little ceremony, that the person who has occasion to use it, does not even interrupt his talk during the operation. I suppose this is one of the reasons why the English ladies, who are exceedingly modest and reserved, always leave the company before the toasts begin.

At last we proceed to the drawing-room, where tea and coffee abound, and where the ladies do the honours of the table with much grace and ceremony; the tea is always excellent, but it is not so with the coffee. Now, since the coffee is not good in a house like this, where no expense is spared, and where I presume it is not brought, already roasted and ground, from the nearest town, in which it is sold by privileged shops, as in London, it cannot be expected to be good any where else in the country. I should imagine that the English attach no importance to the perfume and flavour of good coffee; for it seems to be all one to them what kind they drink, provided they have four or five cupfuls of it. Their coffee is always weak and bitter, and has completely lost its aromatic odour. Thus they are deprived of an excellent beverage, which would be a thousand times better for their health than tea. Kæmpfer, who resided long in Japan, and who has published some very curious observations upon tea and the shrub that bears it, remarks that it contains something of a narcotic nature.*

After tea those who wished retired to their rooms; those who preferred conversation or music remained in the drawing-room; others went out to walk. At ten o'clock supper was served, and those attended it who pleased. I find that as a rule people eat a great deal more in England than in France. I do not know that they are more healthy for it; I doubt if they are; but this I know, that Dumoulin, one of the most celebrated physicians of Paris, declared that he had never been called in the night to attend any person who had not supped.

I have said that I rose with the sun to study the nature of the country, and to make excursions among the neighbouring hills. Here are some details regarding certain banks of porphyry, which, on account of their position, are truly worthy of the attention of naturalists.—

* "I observed," says Kæmpfer "that the tea leaves contain something narcotic, which occasions a disorder in the animal spirits, and makes those who drink them appear intoxicated: this bad quality is partially corrected by the operation of toasting the leaves, which is repeated by degrees; but it is never radically removed: something capable of affecting the head always remains." *Kæmpfer's History of Japan.*

Banks of Porphyry above a Bank of Limestone

Near a lime-kiln, at the end of the park, and on the road which leads to Dalmally, a quarry which has been opened in the side of an eminence, has in the course of its working laid bare its internal organization, the disposition of its banks, and the different materials which compose it.

This quarry is open to the sky, and as a great quantity of stone has been taken out of it for the building of a mill a short distance off, and for banking up the sides of a little stream, which, after winding through the park, serves to turn the mill, the beds of rock are exposed and afford a fine object for study. The position and development of the materials of the quarry can be examined with the greatest ease.

The upper part consists of a bed of vegetable soil, which, though little more than five inches deep, in spite of this shallowness, supports excellent grass.

To this soil a bed of porphyry, immediately succeeds, having a reddish ground, and a thickness of twelve feet ten inches. This bed is divisible into three beds, of nearly equal dimensions, which in some places have undergone a kind of contraction into curious rhomboidal forms, close to which they are in other places simply divided into irregular longitudinal fissures.

The mass of porphyry, twelve feet thick, rests upon a bed only two feet ten inches thick, consisting of tender and even earthy schist, of which the base is a reddish-yellow ferruginous sand, mixed with argillaceous material, soft to the touch. Though in a state of decomposition, this schist is magnetic. I only use the word schist here in relation to the fissile disposition of this bed, which may quite well be only a detritus of porphyric matter.

This schist, in its turn, covers a bed of limestone, averaging about seventeen feet in thickness, which may be considered as a kind of white marble, similar in grain and texture to the marble which is called *saline.* Its whiteness is not every where equal; for by the side of a very white ground there appear some parts of a whitish grain, covered with parallel streaks of a very dark grey, which recall the appearance of a ribbon.

Of this marble, which will take a fine polish, chimney-pieces and jambs of doors may be made. Its grain is scaly and saccharoid, like the finest marble of Italy; but it is rather harsher and rougher to the touch, though none the less capable of taking a lively polish. It is entirely free from any foreign body, and were it not stained with streaks, more or less grey in tint, it would be considered a very fine marble. It is applied to no other purpose than that of making lime.

When I said that this thick bed of marble succeeds immediately the bed of argillaceous micaceous schist, I should have added, that the upper part of the rock is intermixed with small beds, or rather streaks of micaceous steatite, mingled with the particles of marble. Its hardness is not thereby affected, and it becomes a kind of cipolin. But this mixture of steatite and mica penetrates only about an inch into the marble, which is thereafter perfectly pure. The beds of rock form an obtuse angle towards the centre of the quarry; the left side inclining considerably from south to north, and the right side, from west to east. According to every appearance, this position is the effect of some great movement.

Thus there is incontestably porphyry lying above limestone, modified into marble. The ground of this porphyry is reddish, with numerous crystals of felspar, of a dull white, and some much larger crystals of black schorl, less compact than those of the red porphyry of the ancients, but in general very brilliant. The base of the antique porphyry is much harder than that of the porphyry of Inverary, which is somewhat earthy : but this base is fusible, like that of the antique porphyry. In short, it is in every respect a veritable porphyry.

From these circumstances the quarry of Inverary is very remarkable, and ought to be considered as an object well worthy of the attention of all who have it in their power to visit it.

Time flies quickly away when passed in such good company, and in so agreeable a manner. It requires a certain mental effort to leave with composure, persons whose natural affability is so prepossessing at first sight, and to whom one's attachment increases in proportion as they are better known.

Alas! what is life but a series of continual privations?—Let us add this one to so many others; but let us learn to perpetuate our enjoyments by remembrance, and above all by gratitude. Let us leave this delightful mansion; but never let us forget its worthy inmates.

CHAPTER XI

Departure from Inverary.—Arrival at Dalmally. — Scotch Highlanders, their Dress.—Ancient Tombs.—Patrick Fraser. —Reception at the House of Mac-Nab, who possesses several Fragments of the Poems of Ossian.—Manner in which the Habitations of the Highlanders are warmed and lighted; their Usages.—The Circle or Altar of the Druids

WE left Inverary Castle at half past eight in the morning of the 17th of September, by the road to Dalmally. The contrast now experienced was striking; for scarcely had we lost sight of the most charming abode, and the most amiable hosts, when we entered a chain of arid mountains of the most savage aspect.

The road is narrow, rough with blocks of porphyry, and bordered in many places with banks of black schist, cut perpendicularly, the lugubrious colour of which, intersected with large veins of white calcareous spar, seems to announce to passengers the mourning of nature in this frightful solitude.

This melancholy and painful road, where during eight hours we met with no living creature, neither habitations, trees nor verdure, is alike fatiguing to the body and to the imagination. Our very horses, though fresh and well rested, seemed to be tired of it.

More than once I asked our guides whether this disagreeable journey would at last come to an end, when, about four o'clock in the afternoon, one of them shewed me a small column of smoke at a distance, and said "that is where we are to sleep, and we'll be there in half an hour." In a quarter of an hour after, indeed, we found ourselves out of the kind of prison, in which we had been, as it were, shut up since the morning.

A pretty valley [Glen Orchy], dotted all over with hillocks, suddenly opened before us. A small river, called *Orchy*, wound along on the left, several houses grouped together, others scattered here and there, a chapel in the bottom, and a lake in the distance, embellished the landscape. This place is called *Dalmally*.

The inn is conspicuously situated on an isolated eminence, surrounded with verdure. On our arrival there, we saw some fifteen Highlanders in front of the door. All of them wore the same remarkable garb, and they saluted us in a very polite, but at the same time, somewhat haughty manner. We could comprehend from their gestures that we were the subject of their conversation; for they only spoke the Celtic language. But our host, who received us with a very good grace, and who knew a little English, said that we might make ourselves easy with respect to these men, who, little accustomed to see strangers in so retired a place, where few passed, were fixing their attention upon us with pleasure.

"You may be assured," said he, "that these good mountaineers, far from doing you any injury, would, on the contrary, think themselves happy to be able to practise towards you the laws of hospitality, which they love, and which have been respected here from the remotest times. If you find them gathered together in so large a number, it is because this is the Sabbath."

We knew, indeed, that the Scotch Highlanders, being very zealous presbyterians, are rigid observers of religious worship, and will not indulge themselves on that day in the most trifling amusement. They had just come from sermon, and were resting themselves a little before retiring to their respective homes. Their grave and reflective air formed a singular contrast with the gaiety and showy colours of their military attire.

Their costume is very singular. It consists of a military jacket with collar and cuffs, made of woollen cloth, with large checks, red, green, blue, and white; of a large cloak of the same stuff, tucked up and fastened on the left shoulder, called the plaid; and of a kind of petticoat, short and plaited like the lower part of the military dress of the Romans, which serves them instead of breeches; but does not come down below the middle of the thigh. Their legs are thus partly naked, being covered only with woollen half-stockings, of bright colours, in cross bars so that they may closely resemble the ancient buskin. Their head is covered with a blue bonnet, surrounded with a narrow border of red, blue, and green, and decorated with one long and waving feather. They always have a poniard, and frequently a pair of pistols, in their girdle; this poniard bears the name of *durk* or *dirk*. Their shoes, which, in general, they know how to make themselves in a rather coarse but stout fashion, are tied with straps of leather; this kind of shoes is known by the name of *brogues*.

Their money is kept in a waist-belt of otter-skin, which serves at the same time as an ornament. It is so formed that the skin of the animal's head is placed in front. The eyes are edged with a ribbon of red wool, and the head, surrounded with a great many small strings of different colours, covers a pouch, which serves by way of purse.

Such is the attire which the Scottish Highlanders, as well as the inhabitants of the Hebrides, have worn from a very remote antiquity.* Did they copy it from the Romans at the time when those lords of the world attempted in vain to conquer them; or have they received it by a more ancient

* According to Diodorus Siculus, the ancient Celtiberians also wore parti-coloured garments. "*Vestibus utuntur mirificis, tunicas nempe tinctas, et variis coloribus floridas, quasi illi gestant*" Diodor. Sic. V. It should be remarked, that the Scottish Highlanders and Hebridians still speak the Celtic language, that of the Celtiberians.

derivation from their ancestors, the Celts? This question is not easy to answer.

What is quite certain is that the modern descendants of the ancient Caledonians are so attached to this form of dress, which reminds them of their ancient valour and independence, that the English government, having repeatedly attempted to induce them to lay it aside, have never been able to succeed: though this attire is certainly least adapted to a people who live in so cold and humid a climate as this.

After taking leave of our pious Highlanders, we went to see our lodging, and were astonished at its elegance in so desert a place. The threshold of the door, as well as the stairs, were scoured and strewn with fine shining sand; the dining-room was covered with a carpet; the beds were clean and good: and the landlord was a worthy man.

We informed him of the object of our journey; and as we were now only one day's travel from the place of embarkation, we asked him whether it was possible to find here a guide speaking the language of the country, and at the same time, knowing a little English. "Gentlemen," said our host, with an air of eagerness, "I shall be able to arrange your business, and to procure you a man who knows several languages, and will serve you as interpreter and even as guide; for he has already visited several of the isles where you intend to go. He is one of my best friends. The only thing to be found out is whether his employment here will admit of his being absent.—I am going this instant to enquire."

So saying he started off with an agility and vivacity, which astonished me, in a man of his age. I never met with anyone so obliging and so gay as this good Scot.

He came back a few minutes after, bringing with him a man of about twenty-eight years of age, of modest and gentle manners, whom he presented to us under the name of Patrick Fraser, school-master of Dalmally. We soon made our acquaintance with him. This man had prosecuted his studies with advantage, at the University of Edinburgh; he knew very well Latin and Greek, spoke English, and was completely master of his native language, the Celtic, which he regarded as one of the richest and most harmonious.

Patrick Fraser was passionately fond of

the verses of Ossian; and often made excursions among the inhabitants of the mountains in quest of further fragments of these ancient poems. He had already procured as many as would form a considerable addition to the collection of Macpherson, and often made lucky discoveries in that respect, from the extreme pains which he took for the purpose. The narrowness of his means compelled him to exercise the humble occupation of a school-master.

The children of the place assembled in a kind of hut, constructed of dry stones [that is, without mortar]. Here the poor Patrick Fraser taught them to spell the Celtic, or Erse words, printed in the common character; for it would appear that the original characters of this language are lost.* I felt for

* Knox seems to have a different opinion; and as what he says on the subject is well worthy of attention, I extract the following passage from his book: "To these observations I shall add a few facts, to prove that we had for a long time back a written language. In the island of Mull, in the neighbourhood of Iona, there has been from time immemorial, till very lately, a succession of *Ollas*, or *Graduate Doctors*, in a family of the name of Maclean, whose writings, to the amount of a large chest full, were all wrote in Gaelic. What remained of this treasure, was not many years ago bought up as a literary curiosity, at the desire of the Duke of Chandos, and is said to have perished in the wreck of that nobleman's fortune.

"Lord Kames (Sketches, b. i.) mentions a Gaelic manu-

the condition of this estimable and modest man; and when I expressed my surprise and astonishment at seeing him reduced to follow such a vocation for subsistence—"I console myself," said he, mildly, "with my studies, and with the desire of extending my information. It is true, that I sometimes feel uneasy, when I reflect that here I am destitute of every source of instruction. I should doubtless prefer living, though only on a little bread and water, in a city where I could find the means of gratifying my tastes;—but one must learn to accommodate oneself to circumstances."

I should never have expected to meet with a philosopher of this kind, in such a

script of the first four books of Fingal, which the translator of Ossian found in the isle of Skye, of as old a date as the year 1403. Just now I have in my possession a mutilated Treatise of Physic, and another of Anatomy, with part of a Calendar, belonging probably to some ancient monastery, all in this language and character. These pieces, when compared with others of a later date, appear to be several centuries old. I had the use of another equally ancient, from Capt. M'Lauchlan, of the 55th regiment. From these observations and facts, it clearly appears, that ever since the time of the Druids, the Gaelic has been always a written language." Knox's *Tour through the Highlands of Scotland*, &c., vol. i.

It is to be regretted that Knox has not said a word respecting the form of the characters which they used. [The extant Highland manuscripts are written in characters like those of Ireland, some of which go back to the 7th century.]

place. A thousand times did I regret that it was not in my power to charge myself with the future fortunes of a man so interesting and so uncommon.

I stated to him our wish to receive information from him respecting the manners and customs of a country so familiar to him; and I asked whether he could give us the pleasure of accompanying us to the isles of Mull and Staffa.

"Most willingly," said he, "but I ought first to obtain the consent of the parents whose children I teach. It would be inconsiderate in me to ask for a longer space of absence than eight days; if they consent to this, I am at your command." We begged him to do us the favour of supping with us; and as it was yet early in the evening, we engaged him to go and speak with the principal inhabitants of the place, on the subject of the requisite permission. He took leave of us, and went to accomplish this object.

I took advantage of the rest of the day, to make a natural history excursion around Dalmally. I found that the stones which enter into the composition of the mountains here, are, in general, micaceous argillaceous schists, of a greyish or blackish colour, a somewhat fibrous texture, and separating into leaves, more or less thick. The elements of this stone are pulverulent felspar, quartzose earth, clay, mica, and a little iron.

I likewise examined the stones, which have been swept down, and rolled along by the torrent of Glen Orchy. They consist of blocks of granite, massive black spathic schorl, and of compact lavas of the nature of basalt. All these varieties of stone are rolled, their angles have been rounded off; the torrent carries them from a distance.

I was drawn by a great smoke towards the slope of a hill, where I presumed they were burning lime. But as I had no suspicion of limestone here, and was afraid of being deceived, in order to settle the matter I walked to the place whence the smoke proceeded, though it was some way off.

I saw there, indeed, a lime-kiln at the foot of a quarry, with a vertical face behind. I attentively observed the order and disposition of the substances here exposed; of which the following is an account:

1st, The vicinity of the quarry, and, in

general, all the surrounding hills are composed of fibrous micaceous schists, which do not effervesce with acids.

2dly, The quarry is covered with a layer of quartzose sand, of a yellowish colour, and three inches thick.

3dly, The sand is succeeded, particularly towards the left part of the quarry, as one views it in front, by very thin layers of micaceous schist, of the same nature with that of the neighbouring hills.

4thly, These thin layers of micaceous schist lie contiguous to two beds of limestone, or rather of calcareous spar, white, compact, hard, grained like salt, and intermixed with small scales of silvery mica, dispersed through the calcareous paste. Each of these two beds is four feet thick.

Here then, in a space of fifteen miles, there are two instances, nearly alike, of limestone interposed between rocks of a different kind: that of Inverary, where the calcareous strata, intermixed with mica, lie between beds of porphyry; and this of Dalmally, where the limestone is covered with micaceous schist.*

*** [To understand the interest of this observation to the author, we must remember that in his time, a definite succession of rocks**

But it is important to observe, that in both quarries the calcareous matter is in the state of spar; that is to say, it has undergone a confused crystallization, during which tumultuous and rapid operation, it has seized some particles, or rather small scales, of mica. It is natural that in these circumstances, traces of marine bodies are never found, as the calcareous matter is here in a sparry state, that is, it has been held in a state of solution so that the first state in which it existed has been effaced by the displacement and suspension of its particles in a fluid which has given rise to a confused crystallization.

I do not think, however, that the opinion would be well founded, that the limestone, in which no trace of organic bodies can be detected, is the product of an original earth, formed by nature without the aid of shells, madrepores, or other marine bodies. For how can we be certain that the calcareous strata here mentioned, have not been formed at a more early period by the animals of the ancient ocean, whose distinctive characters have been

had been formulated by Werner as universal formations everywhere following the same order. Faujas hàd here found two cases in which the order was not the same, and in which both differed from the sequence proclaimed by Werner.]

subsequently effaced by solution, by displacement, and by a secondary crystallization.*

We see every day instances of new aggregations, destructive of the primitive forms. The grottos of Antiparos, of Notre Dame de la Balme, and so many others, exist in the midst of calcareous rocks full of shells. The daily drippings in subterranean caverns form considerable masses of stalactite from the roofs of the vaults, and of stalagmite on the sides, and bottom. Should we be justified if, on examining tables made of these stalagmites, these calcareous alabasters susceptible of so fine a polish, and having a semi-transparency so agreeable to the eye, we were to pronounce that, as they exhibit no vestige of organised bodies, these secondary rocks are the products of a primitive calcareous matter, that is, of material which came in this con-

* [This sagacious reasoning, following the earlier deductions of Hutton, has been amply supported in recent years, by the detailed researches of the Geological Survey in the Scottish Highlands. The Argyllshire limestones have been traced across the Grampian mountains to the coast of Banffshire, and have been everywhere found to lie in the midst of a series of altered sedimentary rocks. They have not yielded any fossils, their crystalline condition having destroyed their original structure, but traces of worm-burrows have been detected in the associated quartzites, which, from their chemical constitution, have been better able to withstand metamorphism.]

dition from the hands of nature? As if it were allowable in good logic thus to break the thread of analogies; as if a series of analogies did not at least amount to probabilities; as if the expressions, primitive mountains, original calcareous matter, were not unintelligible abstractions, seeing that we can recognise visible and palpable agents which produce, which even assimilate, if the expression be preferred, the calcareous matter in organic forms. These forms are so many points of recognition, interesting types, fitted to direct us in the painful and difficult path of the history of the revolutions of the earth. Whenever, then, we find them effaced, in circumstances where this calcareous matter, losing its organic forms, has been altered into spar, into alabaster, into stalactite, into saccharoid marble, or even into gypsum, I do not believe that we are justified in pronouncing decisively that these materials are not the product of organic beings, or that they have never passed through the filters of animals. It is the same with the mountains denominated primitive, with those of granite, for example, formed by the aggregation of several substances of diverse origin which necessarily presuppose an existence previous to that of their aggregation in the form of felspar, mica, schorl, quartz, and of calcareous and sometimes ferruginous particles. But I shall proceed no farther, for I feel that this is not the place to discuss these great and important questions.

As the day was now drawing to a close, I was obliged to suspend my researches, and I returned to the inn to rejoin my companions. There I found Patrick Fraser, who told me that he was at our service, and that he had permission to remain ten days with us; which gave us great pleasure.

We sat down to table. Our supper consisted of two dishes of fine game, the one of heathcock, the other of woodcock, a cream, fresh butter, cheese of the country, a pot of preserved *Vaccinium* [bilberries], a wild fruit which grows on the mountains, and port wine—all served up together. It was truly a luxurious repast for the country.

Our host, who was naturally polite, and who besides took us for great personages, because he had seen us arrive with three carriages and four servants, wished to pay us, notwithstanding our objections, all the honours which he shews the Scotch *lairds*. So he put the dishes on the table himself, and took his place behind one of us, in order the better to receive and execute our orders. But wishing to treat him on a more friendly footing, we begged that he would take a seat at table with us. He refused; nor would he accept any thing but a simple glass of wine to drink our health. After supper he sent us some excellent rum and tea, of which we had much need to refresh us. Of these he did partake with us, and we spent the rest of the evening in an agreeable conversation with this worthy man, who gave us some interesting information respecting the manners and customs of the country.

This innkeeper is in easy circumstances, and carefully brings up his children, who are numerous. Patrick Fraser serves as their tutor, while a teacher of fencing, and at the same time of dancing, comes every year from a distance to spend some months at the house, exclusively occupied in giving them lessons.*

* "A commodious inn at this place," says Knox, "is rented at 6*l.* and the window tax amounts to 4*l.* 10*s.* This disproportion arises from the well-judged munificence of the proprietor, who thus, almost at his own expense, accommodates travellers with decent lodgings."

At sunrise of the next day, I went to visit the church, which had attracted my attention the preceding evening by the number of tombstones in an adjoining field.

This country church is modern; but is in a ruinous condition, both without and within, which indicates the poverty of the inhabitants of the place.

But a large number of tombs, and some ruins, on a level with the ground at a little distance from the church, show that there formerly stood on that spot some very old religious monuments, probably destroyed at a remote period: for if fanaticism had come round here, in its usual fashion, at the time of the Reformation, we should have found more considerable remains of buildings, nor would so many tombs have been respected at a time when they respected nothing.

The most ancient of these tombs, looked at from the side of art, present a strange character, so remarkable, that, for my own instruction, I thought I ought to examine them very carefully.

All of them are constructed of *lapis ollaris*, or soft serpentine, of a deep grey colour, nearly resembling the stone of which the

Castle of Inverary is built. This stone has completely resisted the action of the weather, and the tombs, which appear the most ancient, are in fine preservation. Their form is that of a simple sepulchral stone, somewhat narrower towards the bottom than the top, like coffins. But I know not whether they have always been in their present state, or whether they have not formerly served as covers to tombs, hollowed out like troughs, such as most of the Roman tombs were. *

Several are cut into right angled parallelograms. Some are five feet eight inches, while others, which are the most numerous, are only five feet three inches in length.

* [The kind of stone referred to in the text, was largely used for sepulchral monuments during the time of the Celtic church in Scotland, and many examples of it, belonging to successive periods of Celtic art, are to be met with among the ancient graveyards of the South-western Highlands. It has sometimes been affirmed that these monuments were manufactured by the monks of Iona, who supplied them, no doubt for a consideration, to the families of the chiefs on the mainland. Another assertion has found some credence that the stones were carried off at a later time, by force or otherwise, from the cemetery of Iona. But whether or not they were the work of the Iona ecclesiastics, there is no such peculiar stone to be found in that island, so that if the stones were carved there, they must first have been brought from a distance. The peculiar schist which furnishes such soft yet durable slabs is found at various places in Argyllshire, but the actual locality whence probably most of the stones were obtained, has not yet been identified.]

A single one measures only three feet and a half in length, and eight inches in breadth; it is probably the tomb of an infant. The rest are, in general, fifteen inches broad. All these monuments are placed flat on the ground, in various directions.

Among these sepulchral stones I counted eighteen, scattered here and there, which from their workmanship, seemed to be the most ancient. They have neither inscription nor date, but they possess a peculiar character in being all overloaded with sculptures in demi-relievo, figures of warriors armed with lances, bucklers, poniards, and arrows, while a bonnet in the form of a mitre covers their head.

On others are seen figures of common horses, among quadrupeds of an odd and fantastic shape, placed beside each other, as on hieroglyphic stones, and in the midst of all, combatants. These bas-reliefs are generally enclosed within a border, which is itself of a peculiar character, for it consists of arabesques, loaded with interlaced ornament, resembling filagree work, in which the strands cross each other in so many different directions, that it is impossible to form a clear idea of what they were meant to represent.

It is superfluous to observe, that the figures of men and animals are executed without any regard to measure, proportion, ground, or perspective, but nevertheless are expressed with a kind of neatness. The character, the physiognomy, if I may use the term, of these monuments, have no connection with the Gothic style; and, save for the border, which somewhat resembles Turkish ornaments, I know nothing with which they can be better compared than with the hieroglyphic figures.

It would be difficult, therefore, to assign the epoch when these monuments were erected; for their style of art has so peculiar a character, that the comparison of it with any thing we already know, would not supply us with a guide.

Some of the natives believe in the tradition, that the stones mark the sepulchres of celebrated warriors who lived in the times of the old kings of Scotland. But the history of these kings, is itself full of obscurity or fable. Others suppose that these tombs contain the remains of the heroes of the North, at the epoch when the Danes made frequent incursions upon the coasts of Scotland, of which they gained possession at different times.

But does it not appear very extraordinary that either the one or the other should have chosen for their place of interment a wild country, buried among mountains, almost uninhabited, at the distance of a day's journey from the sea, and without any road leading to it? What motive, then, could have determined this preference?

Would it not be more probable to suppose, that this burying-place had been formed in consequence of some great battle near it? But the number of these tomb-stones, the long and difficult labour requisite to carve them, must necessarily imply a degree of time and leisure, and must be held to indicate that they were the work of a settled people, to whom the arts, however little advanced among them, were nevertheless not strangers.

These tombs, then, are certainly deserving of the attention of the learned society established a few years ago, at Edinburgh, for inquiring into the ancient monuments of Scotland, and of another lately formed for the same purpose at Perth. I invite them, by their love of the sciences and their country, to be pleased to take up the consideration of this subject.*

In the mean time, those who are curious to form an opinion of the style of this kind of monuments, may see a representation of one, found in a different part of Scotland, in a work of Mr Gardiner, minister of the gospel at Banff, entitled, "Antiquities and Views of the North of Scotland; London, 1780," in small 4to, page 64. This book is written in English, and embellished with very good engravings.

* If some facts stated by Knox respecting the antiquities of the Isle of *Iona*, or *Icolumkill*, one of the Hebrides, are correct, as there is reason to believe them to be, the ancient monuments above mentioned might indeed have been transported thence. But it is still difficult to believe that they should have been carried so very far inland. "This place (Icolumkill)," says Knox, "became also the sepulture of forty-eight kings of Scotland, eight of Norway, four of Ireland, besides the chiefs of the Highland and Hebridian clans, some of whose effigies still remain on the spot; many have been destroyed, and others have been purloined for *other church-yards in the Highlands*.

"I have seen some of these effigies, and also some of the stone crosses, that have been taken from Icolumkill. One of the latter stands in the centre of the town of Campbeltown; a beautiful pillar, ornamented with foliage. The effigies have been carried mostly to Argyllshire, where they are laid over the graves of the principal inhabitants. A number of them may be seen at Kilmartin, where the people can give the names of the persons on whose graves they were originally placed." Knox's *Tour*, vol. i.

Before leaving the tombs of Dalmally, which have, perhaps, detained me too long already, I should like to advert to a circumstance that may show to those who are attached to the arts, and make them their particular study, a kind of uninterrupted filiation which is rather remarkable, though what I am about to mention supposes art to exist now only in its infancy. It will, however, illustrate how much example naturally inclines men to imitation.

The inhabitants of Dalmally having had these ancient monuments before them from generation to generation, have not only preserved the same ground as a place of interment for their dead, but have sought, at the same time, to ornament with sculptures the tombstones which they erect; not, indeed, in bas-relief, which might have been too difficult, but by hollowing out the stone.

The mode of ornament presented by the ancient tombs appearing to them too complicated and meaningless, they have chosen rather, and with reason, to engrave figures connected with their religion, or emblematical of their possessions; and as the lapis ollaris, which they use, is not very hard, and can readily be worked, difficulty of execution has never discouraged them.

We accordingly saw one of these stones, which looked more modern than the others, and could hardly have lain there more than five hundred years, presenting the form of a cross which filled its whole length, but without ornament or figure of Christ. This cross, cut into the stone, is executed with the greatest precision. On others progressively less ancient, are sculptured hammers, chisels for cutting stone, anvils, fishing-boats, nets, in short, all the various implements connected with the vocation of the persons deceased. Lastly, the most modern are decorated with hour-glasses, death-heads, or armorial bearings.*

This successive imitation, connected with a religious custom, must necessarily be limited, and cannot have made progress in so wild a country. It is singular, however, that men so poor, so destitute of resources, and so remote from communication, should never have ceased, during so many centuries,

* [The mingling of relics of ancient Celtic art, often of exquisite beauty, with the grotesque and wretchedly executed carvings of post-Reformation time, is a familiar feature among the old graveyards of the western Highlands.]

to have among them sculptors of some sort; and should still have them whilst, in most third rate towns, it would often be difficult to find a man capable of tracing a simple escutcheon, or cipher, on such monuments.

Patrick Fraser came to call me away from my tombs and my reveries, and to announce that we had to visit a very interesting person, who was the possessor of monuments of another kind.

"We have only a quarter of an hour's walk to his house," said Patrick Fraser. "The name of this man is Macnab. He has in his possession a valuable manuscript, containing several of the poems of Ossian in the Celtic language. You shall have the pleasure," added he, "of hearing him sing them; for the recital of the sublime verses of this ancient poet has always been a sort of singing, which the inhabitants of the mountains and those of the Hebrides have preserved and transmitted from generation to generation."

Macnab's house is situated on the top of the hill on which the inn stands. As we went there, Patrick Fraser said to us, "I should inform you beforehand, that the kind of bard whom we are going to see, is at once a lock-smith, black-smith, armourer, and edge-tool maker, which makes him very useful, and much esteemed. He is, besides, endowed with sound sense, and a striking native wit, who will please you."

His house stands in the midst of a group of habitations, which form a small hamlet. We entered his workshop, which is neither large nor magnificent. His brother did us the honours of the place, and received us with the most agreeable tokens of politeness and regard. He told us, that his eldest brother had gone from home two days before, and that he would be very sorry to have lost the opportunity of receiving strangers who had come so far to pay him a visit. "I cannot," said he, "have the pleasure of shewing you the treasure, of which we have been in possession in our family for more than four hundred years, because my brother keeps it under lock and key. But if you can remain here during the rest of the week only, he will be at your command, as he must be home in three days hence. He will recite to you the verses of Ossian, and tell you some very curious things about that great poet. As for myself, I can have the pleasure only of shewing you the buckler of my great-great-grandfather."

He accordingly brought us, a moment after, a large buckler, of a round form, made of wood, overlain with leather, studded with small brass nails, and adorned in the centre with a rosette also of brass. This buckler, known in the language of the country by the name of *target*, was solidly and neatly made, but somewhat decayed with age. This good man, whose expressions were literally interpreted to us by Patrick Fraser, then went with eagerness to fetch several pieces of iron armour, found a few years before in the ruins of an old castle in the neighbourhood. It was nowise different from the armour of the fourteenth century.

While in Macnab's workshop, where his brother received us, I cast my eyes by chance on a poniard, which struck me on account of its elegant shape. Its handle was made of wood only, but of a very hard kind; it was sculptured in a style that left nothing to be desired in the perfection, finish and taste of the workmanship. The sculpture consisted of an interlaced pattern, the threads joined into knots, and passing and repassing each other in the most graceful manner, and without the smallest confusion.

It may be presumed that the first model of this poniard, and also of the dress of the Highlanders and Hebridians, came from the Romans, with whom they were long at war. For the *plaid*, or cloak; the short jacket; the *fillibeg*, or kilt; the *durk*, or poignard; the *target* or shield—were part of the accoutrement of a Roman soldier. I must repeat, that some very powerful motives must have attached them for ages to this mode of dress, for it is incontestably the worst adapted to a climate so rainy, and where the winters are so long.

I asked Macnab's brother whether he would sell me the poniard, for which I said I would give him a good price. "It is not ours," replied he; "it belongs to one of our friends in the mountains, who has received it from his forefathers and would not part with it for any price. He has left it for repair with my brother, who is able to make you one in every respect similar, should you wish it." Upon that, he pulled out a drawer, in which there lay several in an unfinished state. "We never vary," said he, "from this shape, which is a very good one, being agreeable to the eye, and affording, at the same time, a solid hold to the hand which uses it. All the weapons of this kind which are made here, or in the neighbouring mountains, are like these, and that from time immemorial." *

Macnab, after receiving us so politely in his brother's house, requested, with great importunity, that we would accompany him to his own, where we were expected; assuring us, at the same time, that such a favour would do him a great deal of honour among the inhabitants of the place. We yielded, with pleasure, to his invitation.

The cottage, or rather kind of hut, which

* Mr T. F. Hill, who travelled in the mountains of Scotland with the intention of seeing the country, and at the same time of making critical researches respecting Macpherson's translation of Ossian, published a pamphlet, of which he kindly gave me a copy, when he was on a journey in France. This publication, which contains only some thirty-six pages of print, is entitled, *Ancient Erse Poems, collected by T. F. Hill, among the Scottish Highlands, in order to illustrate the Ossian of Macpherson.* This traveller, in going past Dalmally, did not omit to call upon Macnab, who boasted to him that all his ancestors, for four hundred years back, had been blacksmiths. Mr Hill says that he was far from wanting sense and information; and that he is placed, by Mr Smith, among the authorities for his collection of Erse poems.

*Inside of the Cottage of Mac Nab: a Blacksmith at Dalmally
who Possesses some Fragments of the Poetry of Ossian.*

he inhabits, is sunk several feet in the ground, to be sheltered from the most severe cold; but being placed on an elevated situation, it is safe from damp.

It was supplied with every thing requisite for a comfortable household in such a place. It was divided into two rooms, besides a closet on the ground-floor; for there is no question here of houses with stories; rural architecture is still in its primitive state.

The room to the right of the entrance contained some sacks of barley, and a little oatmeal. These are the only kinds of grain which come to maturity in this country; still they have to be dried in kilns after they have been reaped, towards the middle of the month of October. We saw there, also, some bottles of whiskey, a badly-made kind of spirit, with an empyreumatic taste, which they distil from barley; but it constitutes their best liquor, and is the object of their chief sensuality. We were also shown a rather neat wardrobe, in which was some linen, and the fine dresses in the Roman style, kept for holiday use. The same room, though not very large, and also the closet, contained some simple and plain dairy utensils, and a quantity of peat, very artistically built up against one of the walls. The smallest corners of this little place were put to use, and every thing had its appropriate place. It could be seen that Macnab's brother was a lover of order.

The second apartment seemed to be used as the chamber of reception. Here his relations were assembled, and waited to receive us with ceremony.

A peat fire, on a large round stone, raised ten inches above the floor, and placed in the middle of the room, served to warm it. The smoke rose vertically through an aperture in the middle of the roof. A rustic wainscot, in shape like an inverted mill-hopper, starting at this aperture and gradually widening downwards, descends to within four feet of the ground, at the distance of three feet from the wall of the hut. It is, therefore, necessary to stoop on entering this room, or rather chimney; for it may be truly said, that the salon where the family waited for us, was in the chimney itself. This construction is well adapted to preserve the inmates both from smoke and cold; one feels very warm in this kind of wooden envelope, which retains the heat well. Light enters from the chimney, by two little dormer-windows which have been made in it. A bench, or rather wooden seats were ranged all round the inner part, that is, round the chimney itself. Macnab's relations, who were gravely seated on these benches, rose up as we entered, bowed towards us, and made signs for us to be seated also; they appeared nowise embarrassed. Patrick Fraser acted as our interpreter, and presented our compliments.

When we were seated, a young man shut the window; a second lighted a lamp, of a peculiar form, which gave a large flame, accompanied with a resinous smoke. This economical lamp consisted of a kind of iron shovel, bent towards the bottom into a knee shape, and hung by a long handle in an angle of the chimney, within reach of the spectators. On this were lighted pieces of resinous wood, chiefly cut from the *Pinus tæda*, and well dried, which gave a very bright flame, intermixed, however, with a great deal of smoke. The person in charge of the lamp, has beside him a supply of this wood cut into small bits, with which he constantly replaces what is consumed.*

It was by the glimmering of this extraordinary light, that Macnab, taking by the hand a young, gentle and modest girl, whom I presumed to be his daughter, presented her to us. She carried a very clean wooden bowl [coggie] filled with milk, which she offered to one of us, courtesying at the same time with a little timidity and embarrassment. But, on being encouraged by her father, she drank first herself, according to the custom, and passed it to the person to whom she had before presented it. It then passed from hand to hand, or rather from mouth to mouth, until every one had tasted it, when it returned to Macnab, who closed the ceremony with much solemnity. It should be observed, that we were all standing at this moment. There is in this hospitable custom a sort of religious solemnity, arising from the desire of giving a kind reception to strangers: it is regarded among them as a sacred duty.

* Though wood is now extremely scarce in the country, and there is not a pine of this kind to be seen, nevertheless old stumps of it are found at the depth of several feet in the peat mosses.

We were then presented with butter, cakes made of oatmeal, and a little glass of whiskey. We returned our best thanks to this good family, who insisted on accompanying us back to the inn.* Patrick Fraser informed us that it would be considered as an insult by these obliging people to offer them the most trifling gratuity. This scene appeared to me so interesting, that I made our draughtsman take a sketch of it while we were in the house. (Plate II.)

* The celebrated Johnson visited, in Inverness-shire, a less commodious habitation. His account of it deserves to be copied here.

"Near the way, by the waterside, we espied a cottage. This was the first Highland hut that I had seen; and as our business was with life and manners, we were willing to visit it. To enter a habitation without leave, seems not to be considered here as rudeness or intrusion. The old laws of hospitality still give this licence to a stranger.

"When we entered, we found an old woman boiling goat's-flesh in a kettle. She spoke little English, but we had interpreters at hand; and she was willing enough to display her whole system of economy. She has five children, of which none are yet gone from her. The eldest, a boy of thirteen, and her husband, who is eighty years old, were at work in the wood. Her two next sons were gone to Inverness to buy *meal*, by which oatmeal is always meant. Meal she considered as expensive food, and told us, that in spring, when the goats gave milk, the children could live without it.

"With the true pastoral hospitality, she asked us to sit down and drink whiskey. She is religious, and though the kirk is four miles off, probably eight English miles, she goes thither every Sunday."

On leaving his house, Macnab shewed us, on a small hillock, at a little distance from it, a simple, but very ancient monument, for which the people of the country entertain a sort of veneration; tradition having taught them that these monuments, which bear in their language the name of *carn*, signifying a druidical circle,* were in former times consecrated to the religious ceremonies of the Druids. This one consists of a small circular inclosure, formed by large blocks of unhewn granite.

After examining this kind of altar of rough stones, we prevailed on Macnab and his family to accompany us no farther. But during this contest of politeness, another Highlander came up to us most affably, urgently requesting that we should also pay a short visit to him and his family, who were assembled to receive us.

This man, who was richer and more ostentatious than Macnab, had made his wife put on her best finery, while we were

* [Carn or Cairn means in Gaelic simply a "heap" and is applied more especially to the piles of stones heaped over a grave or where a coffin has rested on the way to burial. The word is also applied to prominent eminences or hills, especially where they have isolated and somewhat conical forms, like vast heaps of stones.]

in the first house. Her toilet, rather hastily put on, though not without some pretension, gave her an amusing air of embarrassment. She came up to inform us, that the fire was lighted, that the table was spread, and that the most excellent whiskey was poured out for us. We excused ourselves, as well as possible, from want of time, shewing her at a distance our carriages with the horses already put to. We then thanked herself, her husband, and the rest of her company, for their obliging offers, and departed.

But scarcely had we gone a hundred paces, when our friend, Patrick Fraser, said to me:—"You have painfully wounded the feelings of that family, who are in easy circumstances, and much respected in the country, by refusing to enter their habitation, while you visited that of Macnab. This kind of preference is regarded among the Highlanders as humiliating." This observation made us retrace our steps in order to repair the involuntary injury. But the woman, on seeing us return, shut the door with a sort of temper, which prevented us from going any farther. We were extremely sorry to find that we had given pain to people so hospitable and so polite.

CHAPTER XII

Departure from Dalmally.—Loch Awe.—Rocks of micaceous Schist.—Porphyries.—Bonawe.—Druidical Monument or Cairn.—Christian Cross in Stone, very ancient.—Night surprises us on the Road.—A violent Storm drenches us.—We lose our Way at Midnight.—One of our Chaises overturned in the bed of a Stream.—Amusing Adventure with an Old Miller.—Reach Oban at half past one in the morning

THE distance from Dalmally to Oban is twenty-four miles; and the road is so bad as to be scarcely practicable for carriages. However, as the day was fine we thought we should be able to sleep at Oban, although it was almost ten o'clock when we started.

We skirted the side of Loch Awe throughout the whole of its length. It is oblong in shape, and more than ten miles long. A road was begun to be opened, at great expense, about halfway up the chain of hills which border this lake. One travels here between two dangers, that of being precipitated from a height of more than four hundred feet into the lake, if a carriage should have the misfortune to be overturned in this narrow road, or that of being crushed by the enormous rocks which fall from the upper steep slopes where such isolated blocks are but slightly attached. This road is really as perilous as it is dreadful; hence it is prudent to take to one's feet at the most dangerous places. The distant prospect and the nearer view are otherwise delightful. This beautiful lake is strewn with little wooded islands. One of these is remarkable for the extensive Gothic ruins of Kilchurn Castle, a second shows a fortress, partly destroyed, and a third is marked by an ancient picturesque chapel. High mountains circumscribe this landscape, and give it a solitary aspect, softened by the beautiful waters of the lake, by the copsewoods which line its slopes, and by these remains of buildings which fill the mind with recollections of ancient times.

In some places, this route resembles that

between Monaco and Genoa, called the *Cornice*. The roadway is cut out of the rock in the same manner. The first mountains, which are the most craggy, are composed of a grey *lapis ollaris*, divided into very thick banks. These magnesian rocks vary in fineness of grain, in colour, and likewise in texture. Some of them are in grain and colour similar to those of which in Italy culinary utensils are made. Others, which are softer to the touch, and of a blackish colour, are susceptible of a fine unctuous polish. A few have a fibrous texture. These beds are crossed at intervals by veins of white quartz.

We proceeded thus for the space of twelve miles in this difficult road, often cut through rock, or constructed among the debris of the mountain side, without meeting with a single cottage; we only saw upon a somewhat wooded slope a few cabins of charcoal-burners.

The nature of the rocks changes as one draws near the thirteenth mile; their colour is a violet-red. Here the mountains seem to recede a little; but their height becomes much more considerable. These mountains are of porphyry, and the road, which there widens, is formed on the detritus of that stone.

On a small arm of Loch Awe, by which it discharges its surplus waters into Loch Etive, there is a bridge at a place called Bonawe. The channel over which this bridge is built, bears the name of Pool Awe. Loch Etive, into which it runs, is an inlet of the sea, stretching up thus far among the mountains, and navigable for large vessels.

The porphyric rocks, which form the neighbouring mountains, have undergone so much waste, whether in consequence of being naturally detached by their own weight, or by the operation of frost, or by other destructive causes, that their base is cumbered with considerable masses of porphyry, broken up into fragments. These heaps have accumulated to such a degree from the weathering of this incoherent rock, that they form little hills at the bottom of the elevated chain. This porphyry, which has a base of trap, and a colour like that of wine-lees, has a great tendency to split up into angular chips of various shapes.

The crystals of felspar, which enter into its composition, are almost all of an irregular form: but what is remarkable, and what particularly deserves the attention of the naturalist, is that in the soil formed of so much detritus, there appear fissures, in which, by the help of the daily infiltrations which keep them wet, the particles of felspar draw together and separate out, in order to re-unite and become aggregated into solid bodies, which have a tendency to crystallization, and present some resemblance to the felspar of Baveno, but have not nearly the same regularity, because this kind of secondary regeneration of felspar takes place here in too rapid a manner. These crystallizations of felspar may be seen in some of the chinks of this moveable soil, at a distance of a few steps from the bridge of Bonawe, upon the escarpment which forms the right bank of the stream that serves as an outlet to the lake.

Somewhat higher, and out of the bed of the river, there are heaps of broken porphyries, which are harder, and have no tendency to decomposition. The crystals of felspar, found here, are in pretty regular parallelopipeds, white in colour, while the base of the porphyry is reddish, sometimes grey. In some parts also the base is yellowish.

Led on by the attraction which this field of observation presented to us, and busy in collecting specimens of these different stones, we did not sufficiently reckon with the time. We had still twelve miles to get over, and we were assured that the road was no better than that we had already passed.

We crossed the bridge of Bonawe, beyond which we saw two solitary houses, inhabited by shepherds; and about a mile farther on, a small inn, by the wayside, surrounded with a few houses, where we found it necessary to stop for half an hour to refresh the horses. While waiting, I walked to a small chapel, on a neighbouring eminence, in the burying-ground of which I observed two sepulchral stones, of the same kind as those I had seen at Dalmally, with sculptures equally ancient. They appeared, however, to have suffered more from the action of the weather.

It would, perhaps, have been more prudent, had we stopped all night at this little hostelry, miserable as it was; for there was no other habitation between it and Oban: but Patrick Fraser assured us that we might go on; that he had once before gone the same journey; that we had still two hours and a half of daylight; and that the moon, which would be shining, would light us for the remainder of the road. We followed his advice and departed.

About a quarter of an hour after we had set out, we observed on a hillock, fronting the road, a cross of black stone, of the nature of slate, upon which a figure of Christ was carved in demi-relievo. The style of the figure was indifferent, but the execution was fine. The figure and the cross were in one piece, and the stone was about five feet high.

We were much astonished to see a religious monument of this kind so well preserved in a protestant country. An old shepherd, who came up while we were looking at it, told us that he had learnt from his parents, that this cross had stood in that place for more than four hundred years; and that although there were no Roman Catholics in the parish, and though all the churches had been destroyed at the time of the Reformation; yet this cross had been allowed to remain, for what reason he could not say, except that the people of the country having been accustomed to see it from father to son, had preserved a kind of respect for it, though they did not pay to it any devotion.

At the same time there was pointed out to us, at a distance of five hundred toises from this cross, a large cross of rough stone, on which, as Patrick Fraser was told, "the Romans had sacrificed to their false gods." Such were the words of an inhabitant of the country, who looked like a school-master, and who spoke a little English. We could not resist our desire for a closer examination of this column, which is not far from the road. It is of a yellowish grey granite. Its shape is somewhat triangular, but it is as nature formed it; for it is impossible to discover upon it the smallest trace of human workmanship. It stands on marshy ground, in a kind of peat-moss. I measured it, and found that its height was ten feet above the ground, and four feet below, which makes its length fourteen French feet. Its mean breadth is about two feet, and its thickness two and a quarter.

At a short distance from this pillar a circular enclosure, about twenty-four feet in circumference, is formed by large blocks of rough granite. This again is one of those very ancient monuments, known by the name "cairn," that is, a Druidical circle.*

We spent an hour and a half in examining it: our curiosity made us forget that we had still a great way to go, and that the sun was getting low. Heavy clouds began to darken the horizon, and as the heat had been stifling during the day, we feared that the weather would turn into a thunder storm. We therefore desired the post-boys to double their

* Let us listen for a moment to Pennant, on the subject of these ancient monuments:

"Take a ride into the country (the Isle of Arran): descend into the valley at the head of the bay; fertile in barley, oats, and peas. See two great stones in form of columns, set erect, but quite rude; these are common to many nations; are frequent in North Wales, where they are called *main-hirion*, *i.e.* tall stones, *meini-gwir*, or men pillars, *lleche*; are frequent in Cornwall, and are also found in other parts of our island: their use is of great antiquity; are mentioned in the Mosaic writings as memorials of the dead, as monuments of friendship, as marks to distinguish places of worship, or of solemn assemblies (1). The northern nations erected them to perpetuate the memory of great actions, such as remarkable duels; of which there are proofs, both in Denmark and Scotland; and the number of stones was proportionable to the number of great men who fell in the fight (2): But they were, besides, erected merely as sepulchral, for persons of rank who had deserved well of their country." Pennant's Tour, 1772. part ii. page 203.

(1) Joshua xxiv. 26.

(2) Wormii. Monum. Dan. 62. 63. Boëthius, Scot. Prisc. et recentes mores, 10.

speed, which they willingly did, as drops of rain were already falling at intervals.

We thus got forward at a good pace for nearly an hour in spite of the bad road; but night came on, and the clouds seemed to press against each other, and we heard the tempest growling in the distance, while vivid flashes of lightning quickly followed each other. The moon was not yet visible, but even had she been risen she would have been obscured with heavy clouds. We jogged along, however, in some degree of security for half an hour more, when a tremendous clap of thunder let loose the storm to burst right above us. The rain soon fell in torrents; the darkness grew more profound, and in a few minutes it was no longer possible to see the road.

Patrick Fraser got out of the chaise, and walking forward before the horses, sought for the track of the road by groping for it with his hands. The horses, terrified by the noise of the torrents, by the blaze of the lightning and by the thunder, moved along with fear, halting at every step. At last our conductors advised us to take to our feet, notwithstanding the rain that was falling in torrents;

they were afraid that we might be overturned, and fall down some precipice, for they found that we had lost our way.

Our resolution was soon taken, and it was time; for we were upon the slopes of very steep rocks. Some supported the chaises, some held back the wheels, while others sought for a practicable track. We moved slowly forward in this way, with much trouble and frequent alarms, not knowing where we were going.

At ten o'clock Patrick Fraser, hearing the noise of the sea, said to us, "We are completely out of our way, there is no longer any doubt of it. I cannot tell where we are. Oban, however, cannot be far off; for we have travelled a long time, and we now hear the sea; it appears that we have got upon some high ground; let us try to get prudently out of this scrape."

At midnight our plight only grew worse. The sea was making a frightful uproar at the foot of the mountain on which we were, which redoubled our caution, and forced us to halt every moment. Such was our situation in this wild region: embarrassed by our horses and carriages, and creeping along steep and slippery slopes, where we had ourselves the greatest difficulty to keep our footing, while the rain continued to fall heavily.

With ceaseless activity Patrick Fraser went always the foremost on the look-out. He came back to say that we must turn round to the left, to avoid falling into the sea; that he believed he had heard a stream about two hundred toises ahead, and that by gaining its bed we might find some outlet from the cliffy ground in which we were involved.

We took this course, and at last, but not without great difficulty, reached the side of a small torrent; but the bank was steep, and the noise of the water announced a deep hollow. It was, however, necessary to cross this difficult track amidst briars and broken stones. The first carriage got through successfully, but the second overturned. It was raised up again without any accident to the horses, and with no further loss to ourselves than the alarm it gave us, and some damage to our portmanteaux. The third escaped better.

Having got to the bed of the torrent, in vain did we try to coast along its banks; even those of us who could best cope with the difficulties of the situation had the water half way up their legs. In about a quarter of an hour the noise of a waterfall, not far off, suddenly stopped our progress. A ray of the moon pierced through a gap in the clouds, and its light revealed a few bushy trees, a small meadow, and some cultivated fields. "We are not far from some habitation," exclaimed Patrick Fraser, "we must call for help to get us out of this abyss."

Wet from head to foot, shivering with cold, and worn out with fatigue, we stood together around our carriages below some tall pines, shouting as loud as we could to any persons who might be within hearing to come to our assistance. This scene appeared to me so ridiculous, that I could not help giving way to a loud burst of laughter. Not one of our party was either hurt or ill; so we determined not to be dejected; but on the contrary, to enliven our talk with some jokes.

William Thornton, who possessed a lively imagination, and was passionately fond of poetry, observed to me that the place in which we were was not without charm; that it was fitted to inspire grand and sombre ideas; and if he could have the good luck to warm himself with a glass of rum, he should feel able to write an ode immediately —"We are," said he, "among the places made famous by the exploits of Fingal. This ground has been trodden by the feet of the immortal Ossian.—His very name calls for the Muses, and my imagination warms!"

Scarcely had he uttered these words in a tone of enthusiasm, when an old man, with bare head and white hair, dressed in a floating drapery of the same colour, appeared to us. "It is Ossian!" cried Thornton, "it is that divine poet himself who hastens to us at the mention of the name of his illustrious father. Let us fall at his feet." But the ghost, without uttering a word, without casting a single look towards us, stalked gravely across the torrent, and disappeared.

"Is it an illusion? Is it a dream?" we all exclaimed; for we had all seen the same object, by the brightness of the moon. We were astonished, and remained in a state of uneasy expectation; when in a few moments we heard the voices of men coming to our assistance. The water-fall proved to be only the sluice of two mills, which had been opened, the white phantom an old miller, who, awakened by our cries, had run in his shirt bareheaded to our assistance; but who, seeing horses and carriages, and hearing men speaking a language which was not his own, went off, without saying a word, to rouse his neighbours. These obliging Highlanders came eagerly to drag ourselves, our horses and our carriages out of the kind of abyss into which we had tumbled. They could not conceive how our carriages could have come down into such a place without being broken to pieces. It required all the address and strength of these athletic men to draw them out, which they did by making a kind of road with pick-axes, and so to say, lifting up the carriages on their shoulders.

They accompanied us to the village of Oban, which was only about five hundred toises distant, and led us to the door of the only inn in the place. They made the landlord rise, who was much surprised to see three carriages, with ten persons, arrive at half past one in the morning, in such a pitiable condition. We showed our gratitude to the good Highlanders who had assisted us in so frank and hospitable a manner. Large fires were lighted to dry us, and after drinking a good deal of tea and some glasses of rum to warm us, we went to bed at four in the morning, and slept till ten. The rest and sleep refreshed us, and save for some slight contusions, and some remaining fatigue, all was forgotten and made good when we rose.

This adventure, though it may appear somewhat romantic, was notwithstanding in all its circumstances as I have related it. I should not have mentioned it, had not two motives induced me; first:—to pay a just homage to my dear fellow travellers, who shared, like worthy naturalists, the fatigues and the dangers of the night, and often laughed at this event, which happily had no disagreeable consequences. Secondly, to be useful to those whose taste for natural history may urge them to visit this little frequented country, by informing them that it is absolutely necessary to set out early from Dalmally to go to Oban; and that if the weather be bad, or any accident delay their progress, they should prudently stop half-way, and sleep at the little inn above the bridge of Bonawe, however bad it may be;

for, between that place and Oban there is no habitation.

Oban is a little hamlet by the sea, consisting of six or seven scattered houses. The sea there abounds in fish, and the herring-fishery as well as that of the salmon, form the principal resource of the place, for the natives only gather a little oats and hardly any barley for the distillation of whiskey. The salmon is dried in peat-smoke, and then cut in pieces and put into barrels, which the Dutch come to buy in order to carry them to Spain and Italy as provision for lent. They catch at Oban salmon which weigh more than a hundred and fifty pounds. When this fish is well smoked and a little salted, the people along the coast, as well as the fishermen, eat it raw as a dainty.

The port of Oban is large and safe, and were not the entrance obstructed by some small rocks, which might be easily removed, it would be capable of receiving a large squadron.* Notwithstanding this advant-

* Speaking of this port, Knox says, "Oban lies in that part of Argyllshire, called Mid-Lorn. It has a good highland country behind, with a freestone quarry, Mull and other islands in front, and is of itself capacious, and sufficiently deep for the largest ships. Without, is the island of Kerera, three miles

age, all the shipping of the place, at the time we arrived, consisted of four small vessels, which had sailed on the herring-fishery, and two wretched boats.

The voyage we had to make from Oban to the Bay of Aros was at least thirty-three miles, in the rapid currents of the strait which separates the Isle of Mull from the cliffy coast of Morven. I did not think it prudent to attempt this passage in such small boats, with herring-fishers who did not understand a word of English; upon a sea, too, which is full of reefs and currents con-

in length; between which and the main land is the sound of Kerera, a good road, through which coasters and fishing vessels generally pass, between the Clyde and the fishing grounds in the North Highlands. A custom-house is already erected, Oban is formed by nature, and by a combination of favourable circumstances, for being a principal harbour, a place of trade, a centrical mart for the South Highlands, and the numerous islands that lie in its vicinity."

Knox adds the following words in a note. "Mr Murdoch Mackenzie was employed by government to survey the west coast of Great Britain, from Cape Wrath to the Bristol Channel; also the coast of Ireland; which he executed with great attention, and much to the satisfaction of the seafaring people of the three kingdoms.

"Speaking of Oban, Mackenzie says, 'In the Sound of Kereray there is a very good anchorage for ships and vessels of any size, and it is a convenient place for vessels that are bound either northward or southward. The best parts to ride in, are in the bay of Oban, and opposite to Oban, near Keraray, and between the Ferry-house of Keraray and Ardnachroik.'"

tinually subject to tempests, and of which both Pennant and Johnson have given so discouraging a picture.

I had just read in the work of the latter the touching episode of the shipwreck and death of that brave youth, so full of promise, Donald Maclean, of Coll, who perished in the short passage from the Isle of Ulva to Inch Kenneth. This narrative impressed me so strongly, that although the sea was then pretty calm, I felt the greatest repugnance to embark in such small and wretched boats as were to be found here, which could only carry four passengers and two rowers.*

I should, however, have overcome this kind of aversion to the voyage, which proceeded less from apprehension of danger, than from the recollection of the discomfort which I suffer at sea, had not our host informed me that they expected, in two days at furthest, the arrival of a fishing smack which was rather stronger than the ordinary open boats. This vessel after taking on

* "To go up the Sound of Mull," says Knox, "even in the most favourable season, is a dangerous experiment for a small open boat, such as Oban can supply."

board some provisions at Oban, was to proceed to the Isle of Skye, which gave me easily an opportunity to land in Mull.

I therefore allowed my companions, who were afraid that bad weather would set in, to depart on board of the two little boats, assuring them that I should sail with the fishing smack, which was so soon expected, and thus would not be long in rejoining them. Meanwhile, I told them, I should employ myself in surveying the mountains in the neighbourhood of Oban, which seemed to me to be very interesting.

It will be seen in the sequel how this kind of foreboding, which arose merely from the caprice of imagination, turned to my advantage. My friends sailed in the two little boats; and it was agreed that they should wait for me at Mr Maclean's, in the Isle of Mull.

I remained, then, alone with a servant in this desert place, at the end of Scotland, among men who spoke a peculiar language, perfectly distinct from English. I could only make myself understood by signs. But necessity begets ingenuity: besides, I knew that I should remain very little in the house.

The mountains, which surrounded me, were so varied, so remarkable, presenting so rich a field of observation, that I proposed to myself to examine them with the greatest attention. The pleasures of instruction and of novelty have so powerful a charm, that they would fully recompense me for some temporary privations, and two or three days ought to pass rapidly away when employed in researches of this kind.

Furnished with my hammers, ink, pens and paper, to write down the observations which I should make on the spot, and taking with me some physical and mineralogical instruments, I set out at daybreak, with a knapsack on my back, accompanied with a domestic, my faithful companion, who carried, on his part, a bottle of wine, and some cold provisions; which, however, we did not taste until after several hours of work.

We then took our frugal, but excellent repast; sometimes seated on the summit of a cliff, sometimes in a sheltered cavern on the shore, where the waves, breaking at our feet, exhibited the spectacle of a raging sea, upon which we rejoiced that we were not embarked.

In the evening I returned to my peaceful habitation, loaded with stones and instructive notes. I spread all my riches upon a table; I put them in order, and even admired them, but I did not regard them with the feelings of a miser; for I amused myself in planning beforehand the distribution of them among my correspondents and friends—and I was happy.

I supped with much pleasure, and sleep soon making my eyes heavy, I gladly went to bed; it was hard, but clean, and fatigue made it into down.

But one can hardly enjoy every happiness at once in this vile world. Will it be believed that music of a kind new to me, but very terrible to my ears, disturbed the repose I so much needed? I had scarcely lain down when a cursed piper came and placed himself under my window. He waited upon me every evening in the passage of the inn, to regale me with an air; he then established himself in front of the house. There was no way of making him stop, and he went on to play this noisy instrument until eleven o'clock, with the wish to be agreeable to me, and to do me a kind of honour, of which I in vain endeavoured to convince him I was unworthy.

On the day of our arrival, this man came before our lodging, walking to and fro with equal steps, and a bold and martial expression of face, deafening us with perpetual repetitions of the most unharmonious sounds, without any air or meaning. At first we took him to be a kind of madman, who earned his livelihood in this way; but Patrick Fraser assured us, that not only was this good Highlander in his senses, but that he had the reputation of being an excellent musician of the Highland school; that his principal intention in playing on this instrument before us was to show his joy at our arrival in a country so seldom visited by foreigners. Touched by this hospitable motive, I was prodigal in my applause, and begged of him to accept some shillings, which he at first refused, and seemed only to receive that he might not displease me. He never played but the same air, if I may give that name to a kind of composition unintelligible to foreigners, but which brings to the recollection of the Highlanders and Hebridians historical events which have the greatest interest for them. As the piper had seen my companions set off, he persuaded himself that I remained behind to hear his music: and thinking, that his concerts would be most agreeable to me in the silence of the night, he continued his serenade under my window till eleven or twelve o'clock. Nothing could induce him to desist. I rose one evening with impatience, and not being able to make myself understood by speech, I took him by the hand to lead him to a distance. He returned immediately, however, as one who disputes a point of politeness, giving me to understand by his gestures, that he was not at all tired, and that he would play all night to please me; and he kept his word. Next day I forced him to accept again a small present, and made signs to him that I did not wish to hear him any more; but he was not to be outdone in civility. That very evening he returned, and made his pipe resound until midnight, playing constantly the same air.*

* "The solace," says Johnson, "which the bagpipe can give, they have long enjoyed; but, among other changes, which the last revolution introduced, the use of the bagpipe begins to be forgotten. Some of the chief families still entertain a piper, whose office was anciently hereditary. Macrimmon was piper to Macleod, and Rankin to Maclean of Col.

"The tunes of the bagpipe are traditional. There has

There are at Oban several species and varieties of very curious rocks, in a space of about eight hundred toises in length, by one thousand or eleven hundred in breadth. This vast collection of different stones, deposited here by nature, in consequence of some great revolution of nature, deserves all the attention of those who love studies connected with the theory of the earth.

A boisterous sea, which beats against and tears down with a kind of rage, the cliffs that serve as its barrier, has disclosed the structure of these hills, which appear to have been heaped one above another by some terrible commotions, and by the action of two elements in ceaseless opposition to each other —fire and water.

The base of these mountains has been so denuded, that their flanks are, in a manner, laid open, which allows the observer to study their structure. He is at first astonished to find so much variety and so much confusion.

been in Sky, beyond all time of memory, a college of pipers, under the direction of Macrimmon, which is not quite extinct. There was another at Mull, superintended by Rankin, which expired about sixteen years ago. To these colleges, while the pipe retained its honour, the students of music repaired for education. I have had my dinner exhilarated by the bagpipe at *Armidale*, at *Dunvegan*, and in *Col.*"

However great his knowledge, he would soon find himself bewildered, if the vestiges of subterranean fires, which are easily recognized, did not enable him to explain this discordant collection, so contrary to the usual course of nature.

I examined and re-examined these numerous different materials, with a kind of obstinacy; and, far from being discouraged by this chaos, I felt an interest in unravelling it. I was animated also by the desire of being useful to those who should visit the same places after me, by fixing their attention on the most remarkable objects, and presenting them with the sketch of a labour which may put it in their power to do better than I have done, without encountering the same difficulties.

CHAPTER XIII

Natural History of the Environs of Oban

I HAVE thought it convenient to adopt the following division for the more orderly arrangement of my observations.*

Calcareous Materials

Though the mountains around Oban are, in general, composed of argillaceous schists, lapis ollaris or steatite, or of traps, porphyries, and compact or porous lavas, and sometimes of a mixture of all these substances united and agglutinated together, yet limestones also occur.

On the beach, at a little distance from the inn, to the right hand, limestone of a black colour may be seen. It is disposed in fissile strata like slate, but is, at the same time, hard, sonorous, and nowise translucid in its fracture. It has a fine grain, and splits into plates from an inch to an inch and a half

* [See note by Editor on page 353.]

thick. The beds, which unite into a kind of thick banks, incline towards the sea at an angle of thirty-five degrees. They are intersected in different directions, sometimes transversely, by veins several inches in depth and thickness, of a very white, hard stone, the grain of which is so fine and close, that, at first view, one is apt to take it for quartz.

All these beds of black fissile stone are calcareous, containing only a twenty-eighth part of alumina, mixed with a very small portion of magnesia. The white veins consist of pure calcspar.

The stone is burnt in a kiln formed on the spot, and yields lime of fairly good quality, but which, to render it more active, is mixed with an equal proportion of a somewhat purer stone, which is found in the isle of Lismore, and is carried in boats to the foot of the kiln. The mixture is effected by calcining them together.

It is of importance to know this beforehand; for the limestone of Lismore, has almost the same colour and the same fissile texture as that of Oban, and as the two stones are placed beside each other at the foot of the kiln, one might be easily led into the error of supposing that they are obtained from the same place. I could not detect in either of them, any vestige of marine bodies.

Argillaceous Schists

Following the shore to the left, and passing below a single house belonging to Mr Campbell, one reaches a great escarpment, in an inclined plane, where the rock is bare for a space of several miles. The sea beats upon this coast so furiously, that it has every where torn and furrowed this kind of natural mole, though formed entirely of hard rock.

This excursion should be made during the ebb-tide; for the beds of rock plunge so rapidly to the sea, the surf is so violent, and the waves rise so high with the least wind, that to visit the place at any other time would be attended with evident danger.

Here prevails almost the same order of things, the same disposition of the fissile beds, the same colour of stone, and the same white veins, as in the limestone quarry above described. But the proportion of the constituent parts is completely reversed. The stone of this vast platform makes but a slight and short effervescence, with nitric acid; lime scarcely forms a twenty-eighth part of its composition; and the rest is a mixture of quartzose and argillaceous earths, with a very small portion of magnesian earth. The white veins, instead of calcareous spar, are of white semi-transparent quartz, which strikes fire copiously with steel, and does not give the least effervescence with acids.

But it deserves to be remarked, that though the disposition of the beds and the white veins is the same as that of the lime-quarry, there are also seen other veins, of a substance like that of the beds themselves, interrupting their course, and crossing them in various directions.

These veins, some of which are more than a foot thick, are themselves divided into a sort of net-work of fissures of retreat which present triangular, quadrangular, and rhomboidal figures. The interstices between these blocks are filled with threads of white quartz, which are sharply defined, on the black ground of the stone.

It may be presumed, that this intersection of the beds owes its first origin to depressions, which have given rise to ruptures; and that the fissures thus produced have been subsequently filled with some stony material which, in its soft state, will itself also have undergone contraction, so as to give rise to such natural mosaic work, to those kinds of *ludus*, which affect a measure of regularity.

Some portions of these veins may be broken off, which are worthy of a place in a cabinet; for these remarkable pieces take sometimes the form of quadrilateral prisms, from seven to eight inches long, and three inches thick. The prisms are themselves formed of a multitude of small rhombs, which seem as if soldered together by some streaks of white quartz.

If one ascends this scarp of schist to a certain height of the mountain against which it reclines, one finds that the same substances again make their appearance. But the fissile beds resume the horizontal direction; the colour of the stone is less black; the base is softer, and the clay still more predominates.

The appearance of the place seems to indicate that the escarpment of which I speak, however considerable in extent, once belonged to the neighbouring mountain; from which it has been detached by some great commotion, or what is still more probable, by the continued action of water, which after having undermined its foundations, will have given rise to this vast landslip which has come down in huge masses.

If one ascends to a height of about forty toises on that part of the mountain which has remained intact, and where the horizontal position of the strata shews that there has been no displacement, one comes upon compact lavas, veritable volcanic products, which crown the whole and form a new order of things, to which I shall return presently. But I ought first to mention the rocks of trap and porphyry, which are as it were, embedded amidst this astonishing assemblage of various substances so heterogeneous in appearance.

Traps and Porphyries

A quarry has been opened two hundred and sixty feet above the level of the sea, on the crest of the mountain, in front of the harbour, and not far from an old wall. Climbing to this point by a steep path, past a group of four or five fishermen's huts, we reach a spot which should be visited for the purpose of observing the disposition of the rocks now to be described.

The quarry referred to presents a bare escarpment, more than forty feet high. The first strata which appear, that is, those which serve as a support to the others, are formed of nearly horizontal layers of a greenish stone, in general hard, somewhat sonorous when struck with a hard body, rather harsh than smooth to the touch, though of a very fine grain. Its external appearance is that of a hard steatite.

But on closer examination, it is found to be a stone of the nature of the traps, which fuses before the blowpipe into a black glass, and is composed of a mixture of schorl (which is the most abundant ingredient), in impalpable particles, with a little quartzose earth, a little clay, and some calcareous earth. The greenish colour is due to iron. The beds vary in thickness; the smallest being one foot six inches, and the largest from seven to eight feet.

Several of the beds, in which the base is rather less coherent, have suffered a certain degree of alteration, which has made their grain friable.

To these banks succeed others a little less greenish in colour, and bordering on dark grey; in the substance of which are seen a multitude of small crystals of felspar, white, hard, and shaped in parallelopipeds.

In proportion as the trap-rock passes into the state of porphyry, by the addition of felspar, the horizontal position of the strata changes, or rather the material no longer assumes the form of strata. It appears, on the contrary, as a large mass, divided by cracks from bottom to top. These fissures, sometimes more than an inch wide and irregular, produce enormous blocks of a longitudinal form, which often adhere to each other so slightly, that, losing their equilibrium, they must fall with a crash to the bottom of the quarry, where they break into a thousand splinters, presenting an excellent choice of specimens, and an interesting object of study to the naturalist.

But it is specially deserving of attention, that the crystals of felspar are not interspersed through every part of the rock. In some parts not one is to be seen, whilst other parts are entirely covered with them. These accidental porphyries, if I may use that expression, do not assume any regular form. They seem to have been scattered at random; and appear in large irregular spots, some of which show a surface of six feet, others less.

Further, these spots, in which the porphyry is so well characterised, cannot be supposed to be produced by blocks of that stone accidentally enveloped in the substance of the trap; for the identity of the base in both, and local appearances, do not admit the slightest doubt upon the subject. It is much more natural to suppose, that at the time when the materials which have formed this rock, were held in solution and suspension in the aqueous fluid, the sorting out of the molecules of the felspar took place, in virtue of the laws of affinity, in the places where the constituent principles of this rock occur, and where they have crystallised after the manner of salts.

Besides those porphyries, which exist in solid masses, a very great variety of others may be observed here, in the form of pebbles, or rolled and rounded blocks, which the sea throws up in such immense quantities upon the beach as to lead to the belief that whole mountains of that material have been destroyed and broken up into fragments by means of some terrible revolution. It will be seen from the following particulars whether what I am here advancing be, or be not, probable.

Lavas and other volcanic Products

Compact lavas, of the nature of basalt, abound around Oban, and they deserve by their position, as well as the materials they have overspread, to engage the attention of all to whom such researches are agreeable and familiar.

These burning currents, vomited by the subterranean fires, have, at very remote periods, flowed along the plateaux, and into the hollows and chasms of the different mountains which skirt this craggy coast. It is thus possible to glean some explanation of the state of the mountains, before the eruptions of these ancient volcanos.

The lavas found here are almost exclusively of the compact kind. The basalt is, in general, very homogeneous; for, with the exception of a few specks of black schorl, which are only rarely met with, no other extraneous substance can be seen in it. This basalt is hard, sonorous, and of a fine black colour.

In some parts, the compact lava forms currents, whilst in others it projects into peaks, in vast pyramids, which seem to have had their birth amidst the most terrible shocks and convulsions, when subterranean fires kindled and melted every thing within the reach of their inconceivable voracity.

Several of these volcanic peaks are split into prismatic divisions, more or less regular, which sometimes present gigantic colonnades, particularly opposite to the isle of *Kerrera.* At other times the prisms are of a smaller size, but more perfectly shaped.

About a mile from Oban, on the road to Dunstaffage, and by the sea-side, is a volcanic eminence, crowned with an old half ruinous castle [Dunolly]. The whole south face of this hill is formed of an assemblage of basaltic balls of small size, but, in general, very round, and scaling off in crusts that fit into each other, as far as the centre, without revealing any body which might have served as a nucleus. The same side of the peak, towards the right when viewed in front, displays a multitude of small, very regular, five or six-sided prisms, the lava of which is decayed. These prisms lose their angles by a sort of natural decomposition, and thus produce round balls, which seem to rise out of the midst of the prisms. I have already referred to similar prisms of a larger size in the neighbourhood of Glasgow.

There is also found in the vicinity of Oban, a porphyric lava, which, notwithstanding the state of fluidity it must have passed through, still preserves its crystals of felspar, of which neither the white colour nor grain have undergone more than a slight degree of alteration. This lava is magnetic, and may be referred to species XX, in the 77th page of the work which I have published under the title of "Minéralogie des Volcans,"* and in which I have described similar lavas of the island of Lipari.

But nothing is more singular than the effects produced by a current of lava on the argillaceous schists, which I but slightly mentioned under that title, when describing

* [This work, in one volume, was published in 1784, the same year in which the author made his journey to Scotland.]

the great escarpment that reclines against a mountain [p. 329].

We must transport ourselves to the top of this mountain, which itself reclines against another of still greater height, that we may see there a current of basaltic lava, which descending from the latter, has come to cover the upper and horizontal plateau of the former, and has then flowed from cascade to cascade down the slope of the mountain that fronts the sea.

The face of this tract evidently proclaims, that at the period when this ancient volcano vomited forth the lava, there already existed here great rents and depressions, seeing that the lava has moulded itself to the surface of the ground, following all the accidental contours of the slopes from the summit to the base of the mountain.

This volcano was submarine; of this it would be easy to give several proofs. But for the sake of brevity, I shall confine myself to that, which may be easily comprehended by the greatest number of readers, even by those who are least conversant with the natural history of volcanos. It is as follows:—

When describing in this chapter the argillaceous schists of the nature of slate, I stopped at their point of contact with the basaltic lava [p. 329]. I there stated, that the colour of the schist became fainter at the height of about fifty toises above the level of the sea, that it was rather grey than black; that argillaceous earth predominated in it; and that the strata, which were highly inclined towards the bottom, were horizontal towards the top, particularly in certain parts.

It is there that the observer may easily find some obvious places, where the beds of schist are pretty thin, and also divided into a great many blocks, the most numerous of which are rhomboidal, others are triangular, or quadrilateral. As these solid pieces, which in the aggregate form beds, easily split up and fall asunder, one may select some which are remarkably regular in form, though their shapes are merely the result of contraction and not of crystallization.

The schists with those shapes are certainly not a volcanic product, though they are covered over by a lava. But it is very singular, that neither their constituent parts, nor even their colour, have suffered the slightest alteration from the burning and molten mass that has overspread them, and has moulded itself upon their surface. I proceed to give a proof of this. These schists are of such a nature, that, on exposing one of the rhomboids, taken from immediately below the lava, to the rather prolonged action of an ordinary fire, it soon assumes the dark red colour of brick. The same effect would have, doubtless, been produced by the boiling lava, had it come in immediate contact with the schists in ordinary circumstances.

It must, therefore, be concluded, that some intermediate body, such as water, diminished the action of the heat; and the most natural inference is, that this volcano was submarine; inasmuch as it has not, in any respect, altered a material so sensible to the operation of fire as this delicate schist, which is so rapidly changed to a red colour by the heat which oxidises the ferruginous particles intermixed in its base.

After having made some experiments respecting the action of ordinary fire on these schists, whilst I was meditating upon this interesting fact, and writing my observations upon a table of this lava, that rested on these same schists, in the place just described, I found that on passing my hand under the table, I could easily pull out as many as I chose of the small rhomboids of schist. The plain reason of this was, that the lava, by contracting its bulk as it cooled, left an empty space several inches high, below the table, which was hardly more than three feet broad, and which adhered on one side to a more considerable mass of lava.

As I examined these small prisms of schistose stone, some of which were three inches high, it occurred to me to put some of them up to the loadstone. I was astonished to find them strongly magnetic in the part next the lava, whilst they were nowise so on the opposite end.

It was natural, after this result, to try to ascertain to what depth the quality of attraction extended. This was effected the more easily, as these small prisms, on being dexterously struck with a hammer, split into slices of from half an inch to three or four lines in thickness. I was accordingly enabled, with the assistance of a very sensitive magnet, to determine that the part of the schist next to the lava, had retained its magnetic power beyond a depth of fourteen lines.

I then made use of strong magnifying glasses, in order to examine whether the attractive parts had received by infiltration any ferruginous particles from the lava, which is itself strongly magnetic. But I could discover nothing to confirm that conjecture. The base of these schists appeared to me absolutely homogeneous, and the same throughout; that is to say, fine-grained, pretty soft to the touch, and without the slightest visible atom of iron.

But if we consider that the black schorls which are found in quartz and the granites, and which are not in the least magnetic in their primitive state, become so by the action of fire, as we may be easily convinced of by heating them in the ordinary fire of our stoves, and as may be seen on a grander scale in the immense quantity of schorl-crystals found on Mount Etna; we must conclude that the base of the schists here under consideration may quite well be partly composed of a pulverulent schorl, the general characteristics of which escape observation from the minuteness of their molecules; and that wherever the heat of the lava has been able to reach them, their attractive property has been developed.

These small prisms of schistose stone may, therefore, be regarded as excellent pyrometers, for determining, with the help of comparative experiments, what must have been the heat of this lava, the effect of which has certainly been enfeebled by that of water. For, had it been otherwise, and had not the volcano been submarine, I repeat that the burning lava would have exerted an action of another kind upon the schists. Instead of making them simply magnetic, which does not require a violent heat, it would have changed them to a brick-red colour; as takes place, as above stated, on exposing them to a rather strong fire. I have made several experiments in my laboratory, which confirm what I have now advanced respecting schists of this kind. I reserve them for a work that will admit the details, which are too minute to find a proper place here.

The only varieties of lava which I was able to find in the environs of Oban, may be classed under the following numbers:

No. 1. Compact basaltic lava, pure, black, hard, without any extraneous body, giving a black glass with the blow-pipe, and disposed in large streams.

No. 2. *Idem* in prisms, a number of which form banks of a great height, on the side of the channel facing the isle of Kerrera.

No. 3. *Idem* in small balls, which exfoliate in proportion to the degree of alteration of the lava.

No. 4. *Idem* with some globules of white calcareous spar, not very abundant in general.

No. 6. Porphyric lava, forming streams, sometimes divided into more or less regular prisms.

No. 7. Porous grey, reddish, or sometimes black lavas, more or less hard, often so friable and so decayed, that they moulder into an earthy dust. These and the lavas containing globules of calcareous spar, are only found in masses of a particular conglomerate, which I shall describe immediately.

From this list it appears that the lavas remaining in their original place present but few varieties, though they exist in the form of streams and enormous masses. This is only what might be expected, seeing that the craters, and the scoriaceous lavas which surround them, have disappeared, so that it is no longer possible to trace the places occupied by those terrible fiery orifices. It should seem, therefore, that after the eruptions of these ancient volcanos, catastrophes of another kind succeeded, which swallowed up these dreadful vents, and scattered to a distance the scoriæ, pumice, ashes, and other materials which the volcanos had projected.

What remains for me to observe respecting a final lithological subject from the environs of Oban, may serve as a supplement to what I have now stated.

Remarkable Conglomerates, forming natural Walls of great Thickness, and of a considerable Height

In my "*Mineralogie des Volcans*," page 334, I established a distinction between breccias and conglomerates, which has appeared to me to be necessary.*

Wherever the fragments of any rock, preserving their angles, are collected together and agglutinated by a natural cement, I give to that aggregate the name of *breccia*.

* [The author is well entitled to the credit of this important distinction, which was published by him in 1784, in the work above mentioned, and which has since been often of service in the reconstruction of ancient geographical conditions.]

But if the stony fragments, on the contrary, have their angles knocked off and worn down, if they are oval, or round, whatever be the nature of the pieces, or of the cement which binds them together, I apply to them the term of *conglomerate* (*poudingue*).

This distinction appeared to me requisite, I repeat, because it serves to discriminate two different modes of origin, and presents to us some instructive characters. Splinters or fragments of stone which have preserved their sharp edges do not indicate that, after being torn from their primitive position, they have been the sport of waves, and subject for a long time to the impetuosity of currents, which might have transported them to a distance, for in that case their angles would have been worn off. Their appearance, therefore, shows that they are not far removed from their original situation; while the stones which have lost their angles, and have assumed shapes in every way similar to those produced by the continued attrition of hard bodies, while rolling and rubbing in every direction against each other, require the supposition of a violent and prolonged action, which must have carried them to a distance, or, at least, have kept them, for a long time, in convulsive agitation.

The traveller sees with astonishment, in the neighbourhood of Oban, vast walls of conglomerate, some of which are more than two hundred feet high, and exceed sixty feet in thickness. These walls stretch along the whole of the coast, from the right side of the harbour, fronting the sea, for a distance of more than three miles.*

This kind of natural rampart forms, in some parts, a barrier, which, for many ages, has resisted the impetuous fury of the waves, on the margin of a sea almost incessantly raging. While in some places it varies in height, in others, it is completely isolated on every side, and exactly resembles the walls of an immense coliseum, which may be traversed in every direction, and examined on every face. For the most part this extraordinary wall reclines against the chain of vertical heights with which the coast is

* [The author here mistook the structure of the ground. The conglomerates do not form a dyke or wall, but lie in approximately horizontal sheets, which extend inland under the overlying pile of lavas. The vertical face which they present to the sea, is due to the denudation of that portion of them which once extended westward across the Sound of Kerrera.]

bordered, and which it adheres to, and, as it were, incrusts. This remarkable conglomerate, bound together by a natural cement of the greatest hardness, has sometimes been shaped into detached peaks, which rise in the form of pyramids, or needles, and suggest the idea that they may be grand monuments, erected by human hands. I declare, that since the time when I began to make natural history my chief occupation and my delight, I have never, in all my many travels, met with an object of this kind which has so much astonished me. The pyramidical rock of St Michel, situated in the middle of the town of Le Puy, in the Velay, is doubtless very extraordinary for its conical shape and its height. But it is entirely composed of lava, and owes its origin to a current of molten material, which burst upwards through the earth, and stood upright upon itself, as it congealed by the action of the cold air.* But how have these

* [This remarkable object, in the scenery of the district, is not a current of lava, but a mass of volcanic agglomerate and other material, which marks one of the volcanic vents of eruption in the Velay. Its isolation in the valley has been brought about by the denudation, which has removed all the rocks that originally surrounded the volcanic chimney.]

peaks at Oban been upraised, composed, as they are, of various materials, cemented to each other? This is a question of difficult solution. Let us see whether the different stones, which form these masses, will enable us to propose at least some probable conjectures upon the subject.

Of the different Stones which enter into the Composition of the Conglomerates of the Environs of Oban

1. White quartzes, sometimes reddish, of great hardness, giving fire with steel, worn and rounded on their surface, rather circular than oval, and varying from the size of a hen's egg to that of a small cannon-ball.

2. Oval and rounded fragments of greenish trap, and of grey trap, resembling the porphyric rock described above. The base of these rolled traps seems to have been little altered.

3. Argillaceous schists, black, hard, and somewhat calcareous, nearly of the same nature as that in the escarpment of which I have spoken.

4. A black, calcareous, and somewhat ar-

gillaceous stone, resembling that of which lime is made at Oban.

I ought to observe here, that this stone, as well as the preceding (No. 3), being much less hard than the others, and disposed in rather thin layers, the rolled fragments of them found in the conglomerate are hardly larger than a small walnut.

5. Porphyry, of a greenish, a grey, and a yellowish ground. The last variety is most plentiful; all possess great hardness; and their parallelopipedal crystals of felspar are opaque, and of a milky white. These pieces of porphyry are all round or oval, generally of the size of one's fist, sometimes even larger. Several of them move the magnetic needle whilst others make no impression upon it.

6. Compact, black, basaltic lava, giving with the blowpipe a black enamel, attracting the magnetic needle. This lava, which is itself magnetic, is, in general, very sound, although sometimes a little altered. The pieces of it are all round or oval.

7. Porphyric lava of the same form. Some pieces have suffered in their base, whilst others are altered, and as it were, rusted. All of them, however, are magnetic.

8. Porous lava, heavy, black, and sometimes reddish, having its vesicles filled with white calcareous spar. This lava is, in general, altered, and a little earthy.

9. Porous lava, grey, black, or reddish, the vesicles of which are empty. These lavas are so altered, that they are friable under the fingers, and fall into a gravelly powder.

All these different stones are, I repeat, rounded or oval, of a greater or less size, in proportion to their different degrees of hardness, huddled together and intermingled without order, and attached to each other and compacted into a mass by a cement so hard, that it is exceedingly difficult to separate them with a hammer, which in general rather breaks than disjoins them.

It is not very easy to determine accurately the nature of the cement (gluten), which has so intimately consolidated these immense masses of different kinds of stone. The narrowest interstices, and the smallest cavities, are so closely filled up with a sort of gravelly sand, formed of a kind of detritus, produced by the decomposition of all the stony substances, which have been kneaded and amalgamated together that it becomes extremely difficult to recognise what it is that binds the whole mass so firmly.

On examining, however, the most decayed parts of this cement, with strong lenses, one sees that the pulverulent debris of the lavas generally predominates in it, and that a very fine and subtle kind of lapidific juice,* has bound the whole together in this intimate way. I threw into aqua fortis some particles of this cement, which I detached with the point of a knife, and found it to consist of a mixture of quartz, and calcareous matter, in which the first predominates.

The more one examines this immense assemblage of stony substances of divers kinds, rounded by attrition, the more one studies the form of these enormous masses, their position in the neighbourhood of lavas, and their physiognomy (if I may be forgiven the expression)—the more are they found to differ from ordinary beds of pebbles, which the waters have accumulated in such great

* [This term, now obsolete, was in frequent use during the seventeenth and eighteenth centuries, especially in reference to the remains of plants or animals imbedded in rocks. These fossils were supposed to have been turned into stone by a lapidifying juice with which Nature was supplied.]

abundance in sundry places. One is, therefore, led to find in them a kind of resemblance to certain eruptions of volcanic mud, in which water, heated to the highest degree of ebullition, comes into co-operation with fire, and the different elastic emanations generated by the subterranean conflagrations (incendies souterrains). This cause may have given rise to those rapid and tumultuous petrifications,* of which the remains of ancient extinct volcanos, so to speak, afford us examples at every step.

I am therefore inclined to ascribe the origin of these astonishing ramparts and huge pyramids of conglomerate, to volcanic eruptions of this kind. We must believe that the sea undergoes furious convulsions, when its bottom is upraised by violent explosions, and by the earthquakes that are produced when water is converted into vapour, in the midst of these terrible combustions (embrasemens). Vast displacements

* [This word is used here in the sense of masses of rock. The passage above is interesting as suggesting that Faujas, in spite of his prolonged study of volcanos and volcanic phenomena, had not quite emancipated himself from the time-honoured error, that these phenomena arise from the combustion of inflammable materials underground.]

of materials must then necessarily result; pebbles and fragments of rocks are brought together and mingled with mud, sand, and volcanic ejected matters of every sort, which serve as cement. Hence solid masses of rock are formed, which can afterwards cohere firmly above each other, owing to the binding influence of a cement, which is all the more tenacious, as it is the product of the two most active solvents known, namely fire and water.

Important deductions might, doubtless, be drawn from such important facts. Scientific men particularly conversant in the natural history of volcanos, will be better able than I am to discern them and to point out their application. Moreover, this is not the place to enter upon such a discussion, however important it may be. I even fear that I may be reproached for having already entered into details which may be thought too minute. But the mountains and the lithology of the environs of Oban presented objects so interesting, from their variety and their position, and which were so little known, that I conceived that naturalists would be in some measure obliged to me for giving them an account of my researches.*

Of the Scots Parsley, or Ligusticum Scoticum

It now remains for me, before taking leave of the mountains of Scotland, to mention a plant, which is in high repute

* [This chapter has a special interest as one of the earliest essays in Scottish geology, and the first in which the rocks of the remarkable district of Lorne were brought into notice. It is also interesting from the evidence it affords of the rudimentary condition of petrography and structural geology, towards the end of the eighteenth century. Faujas was the first who clearly recognised the volcanic nature of some of the rocks around Oban, but he regarded as aqueous deposits others that are as thoroughly lavas as those which were so called by him. He came to the right conclusion, that some of the lavas had been poured out under water, but his grounds for this inference were rather feeble. He missed the important fact that the lavas are intercalated among true sedimentary deposits. He was naturally puzzled by the coarse conglomerates: these still present to-day problems for which no completely satisfactory solution has been found. But some of his difficulties would not have arisen, had he perceived that the conglomerates are not in the form of dykes, but of beds underlying the main series of lavas. He was apparently misled, too, by the terraced escarpments of the Lorne lavas, which in one instance he describes as having descended, in cascade after cascade, from the heights to the shore, the molten material moulding itself upon the irregularities of the slope. He seems to have failed to notice that, while the schists are highly inclined, the overlying conglomerates and volcanic sheets are nearly horizontal, and that the escarpments mark the outcrops of successive lavas piled above each other, up to the tops of the terraced hills around Oban.]

among the natives of the country, as much on account of the virtues which they ascribe to it, as for its culinary uses. This plant is the *Ligusticum Scoticum* [Sea-loveage], which I found growing in great abundance, at Oban and by the sea-side at Inverary.

Robert Sibbald, in his work entitled *Scotia Illustrata*, published at Edinburgh, in 1684, was, I believe, the first who described and gave an engraving of this plant; but his description is indifferent, and incomplete, while the figure is bad.

Plunket has also given a representation of it in his collection, plate xcvi. But it is engraved from an incorrect drawing, which does not present a true likeness of the plant.

This want of a good representation of it, has induced me to give one in the present work, wherein subjects of natural history are so often alluded to. I have resolved to publish it with the greater pleasure, as my intention has received the approbation of three of our most celebrated botanists, Jussieu, Lamarck, and Desfontaines, for whom I entertain as much respect as personal attachment.

The plant has been drawn by Maréchal,

an excellent painter in natural history, and engraved by Sellier, whose talents in this line are well known. (See plate I.).

Ray, to whom Jacob Newton had transmitted some account of the uses of this plant, states, that the latter was informed in the country that the Highlanders of Scotland ate some *Ligusticum* every morning, in the persuasion that it was an antidote which would preserve them from every malady during the day. "*Mihi (inquit) ibi notum est, Scotos montanos, apud quos copiose oritur, quotidie mane eam esitare, quo se tutos esse persuadent toto die a quavis contagione.* Ray, Hist. 447.

Gunner says, in his *Norwegian Flora*, that this plant, and also the common loveage, are given, mixed with salt, to cattle as a preservative against diseases. "*Folia hujus, vel et ligustici levistici, plantæ hortensis, cum sali peccoribus ut remedium preservativum dantur.*" Gunner, Norw. 85.

The most modern botanist who has mentioned the properties of the *Ligusticum* of Scotland, is John Lightfoot, in his *Flora Scotica.* He speaks of it as follows:—

"This plant grows on the rocks by the

sea-side in many places, as on the coast of Fife, between North and South Weems, and below Kinghorn, and frequently in the western islands of Jura, Ilay, Iona, and Skye; in which last it is called by the name of *shunis* or *siunas*, gaulish; and is sometimes eaten raw as a salad, or boiled as greens. The root is reckoned a good carminative. An infusion of the leaves in whey is given to their calves to purge them. The dose is ℥vii." *Lightfoot, Flora Scotica, part II. p.* 205, *tab.* 24.

This is all the information I have been able to gather respecting this plant, which the Scotch Highlanders, and the inhabitants of the Hebrides, regard as a kind of universal panacea. In France, our Angelica was formerly held in pretty much the same estimation; it was placed above every thing: and it ought to be observed, that the *Ligusticum* of Scotland is ranked, by Lamarck, among the Angelicas.

END OF THE FIRST VOLUME

Printed in the United States
By Bookmasters